WHAT'S IN A YEAR

A COUNTRYMAN'S TALE THROUGH THE SEASONS

JOHN MILLER

WHAT'S IN A YEAR
First published in Great Britain in 2025
by: LOTUS BOOKS
An imprint of PARTNERSHIP PUBLISHING

Written By John Miller
Copyright © John Miller 2025

John Miller has asserted his right to be identified as the author of this Work in accordance with the Copyright, Designs and Patents Act 1988.

The editors have ensured a true reflection of the author's tone and voice has remained present throughout.

All rights reserved. No part of this publication may be reproduced, stored in a retrieval system, transmitted, or copied in any form or by any means, electronic, mechanical, photocopying, recording or otherwise, without the prior written permission of the publisher and copyright owner.

A CIP catalogue record for this book is available from the British Library.

ISBN 978-1-915200-79-2

Book Cover Design by: Partnership Publishing
Book Type Set by: Partnership Publishing

Book Published by:
PARTNERSHIP PUBLISHING
Lincolnshire, United Kingdom
www.partnershippublishing.co.uk
Printed in England.

Partnership Publishing is committed to a sustainable future for our business, our readers and our planet. This book is made from paper certified by the Forestry Stewardship Council (FSC), an organisation dedicated to promoting responsible management of forest resources.

WHAT'S IN A YEAR

A COUNTRYMAN'S TALE THROUGH THE SEASONS

JOHN MILLER

Charcoal and crayon sketch of myself as a boy circa 1971
by Tom Coates (1941-2023) PPNEAC, PPRBA, PPPS, RWS, RP

PRAISE FOR JOHN MILLER

Nature writing is often a tricky balancing act between over-dramatisation and over-simplification, and between the highly personal and the dryly scientific. John has wisely chosen a middle way through, a simple 'this is what I have learned', and 'this is why it matters so much'.

Like all good nature writing, What's in a Year is based in hundreds of hours of acutely-described observation. What then goes on to make it so special is his genuine appreciation, his generosity and his humility. Whether he is planning his year's vegetable garden, watching black grouse lekking on a Yorkshire moor or following an uncooperative trout up a Hampshire chalk stream, John sees himself as no more important than the creatures around him that he is describing. In doing so, he is embracing that huge truth about our own insignificance, without which there can be no recovery.

He doesn't have to be right- (I think he may give the excesses of shooting too brief a mention, and every shepherds pie should include carrots) - as there is no definite right. But he is passionate, he is very knowledgeable and he is generous with both. What he has produced is a love letter to the countryside that he is privileged to roam across and to respect.

Above all, What's in a Year is a plea for two sides, who should never have become sides in the first place, to come together and chisel out a future that ensures that all our children, and their children, are educated in, and enriched by, the natural world around them.

That is a noble aim!

Roger Morgan-Grenville
Author of The Return of the Grey Partridge and Across a Waking Land.

"In our faster, more digital, and ever greater excesses of modern life we have undoubtedly lost touch with the countryside we purport to love. In his book What's in a Year, thankfully John Miller is leading us back to slowing down, taking a deep breath, and rekindling our connection to the sights, sounds, and smells of the real countryside. Thank you John."

Tarquin Millington-Drake

Frontiers Travel and author of 'Living with Greys'

• •

John is one of those rare people whose infectious enthusiasm and smile make an impact immediately you meet him. His knowledge of everything in the natural world is quite extraordinary, and when reading this wonderful book that is immediately apparent.

His book is a fountain of knowledge, given with the passion of a true countryman. A man who loves his wildlife, and yet is pragmatic about how the conservation of wildlife habitats should be achieved. He loves his field sports and is passionate about preparing and eating food, whether that be from his garden, rod or gun.

To the true countryman, every season is full of promise, with the added excitement of what is around the corner. Keep turning the pages of What's in a Year and you see the Natural world in a new and a very special way!

David White

Wildlife Photographer and Wiltshire Farmer

CONTENTS

PREFACE		1
APRIL	MRS PHEASANT AND MRS OWL	11
MAY	CONSERVATION IN ACTION	37
JUNE	CARLOGIE & BALLOGIE	55
JULY	WE ARE WHAT WE EAT	73
AUGUST	LAXA AND VANHALLE	83
SEPTEMBER	AND THE ROBIN SINGS AGAIN	95
OCTOBER	WESTERN ISLES	117
NOVEMBER	RED LEGS AND FIELDFARES	139
DECEMBER	PULL OF THE TIDES	161
JANUARY	CULTIVATING NATURE'S BIODIVERSITY	185
FEBRUARY	A WILTSHIRE MCNAB	205
MARCH	THE DAWN CHORUS	221
EPILOGUE		233
PHOTOGRAPHS		244
ABOUT THE AUTHOR		249

To nature and the respect it deserves.

I hope this book inspires you to look a little closer at the world around us. There is always something to see and to nurture.

• •

"There is nothing like looking if you want to find something…
You certainly usually find something, if you look,
but it is not always quite the something you were after."
THE HOBBIT, J. R. R. TOLKIEN

PREFACE

There are three robins in the garden, each staking out their winter territories, a solemn whispering melancholy. Not the bold brazen shout of spring, but a reserved and shy ornithological blues, quietly singing in the end of Summer.

We are very fortunate in the UK to have a huge diversity of birds across the multitude of geological-defined landscapes, meaning there is usually something to be heard even in the depths of a stormy December. Spring and summer songbirds bring the great festivals of song, loud and in your face, but maybe the fitful quieter autumn and winter chants are more poignant as their soloists have the stage all to themselves.

So, little robin staking out your territory, sing your seasonal reminder to me that autumn is approaching – your song may be a little sad, but it is also the harbinger of bonfires, apples, and cold walks. Do you need a Robin to tell you that? Not really, but it has been a clarion call all my life: "September is here!"

"What's in a year" is a seasonal view of what to do and see, month by month: whether to grow, cook, hunt, or just observe. My challenge for you, is to set out what you too can see and do to enrich your life when you are better connected to nature and your local area. To widen your eyes, enrich your nose, and tune your ears to celebrate the year. Be close to and nurture your environment. I may have 16 summers left, maybe more, maybe less, either way, I need to get on with stuff. I am now acutely aware of the clock ticking: every day, every week, every season passing by; and, with a drive and purpose that I hadn't felt before, I feel an innate passion to not just sustain but enrich this wonderful world.

To this very day, my heart still races when I find a bird nest. That sheer excitement when I found my first hedge sparrow nest over 50 years ago is still so clearly there in my mind. Its four mesmerising perfectly formed sky blue miniature treasures nestled in their mossy egg box. Whatever this kindled, it was enduring, and my love affair with nature continues. This affinity to nature and its captivating spell, is one that is open to everyone, and the more of us who hold these experiences close to our hearts, the more options we have available to neutralise the menace bearing down, threatening our past and future treasures. For they are treasures. Ones to be cherished and protected.

Our world is changing at pace, and it feels impossible to keep up. The new becomes the norm before we realise, and that is dangerous as we lose our benchmarks, our grounding in our surroundings.

As a boy, I remember a cart track, what we would call a bridleway today, complete with a blue sign to remind us how we are authorised to travel its fairway. We called it Coves Lane in the village of Kidmore End, north of Reading in the Chilterns. I never met another soul there and it was part of my personal kingdom, a playground of natural history. In the summer the evocative soft and so very gentle purring of turtle doves would tell me I was near the overgrown field maple, just before the ancient elm with its tawny owls and then the oak in the corner of the bullock field with a green woodpecker hole. Any walk in the fields would be accompanied by skylarks, rising in a crescendo to fill the skies, above poppy red and corn marigold yellow polka-dotted barley. Behind our house, in the rough 5-acre paddock, at least one covey of grey partridges lived, and, in the village, house martins competed for the best eaves, while spotted flycatchers nested on the church buttresses. Cuckoos laid their eggs in our hedge sparrow nests, and any bowl of milk was quickly found by hedgehogs. In Autumn I would watch the flocks of birds following the plough, marvelling at the lapwings on their blustery wings. A fox was a rare sight, a badger exceptional, crows and buzzards unusual while squirrels, magpies, and crows were limited to the woods - but there were many more kestrels.

Conservation was a tactical reactive subject, aligned to specific events like the Torrey Canyon oil spill in 1967 between the Scilly Islands and the Cornish mainland. The solution to what is still the UK's worst-ever marine pollution incident now appears primitive: the RAF sent planes from RAF Chivenor and Lossiemouth to bomb the wreck and then drop napalm and kerosene in the hope of burning off the spewing oil. Largely unsuccessful and most likely adding to the pollution, this was followed by tugboats spraying thousands of gallons of toxic dispersal detergents. To be fair, one of the consequences of the oil spill has been the improvement of maritime charts and double-skinned hulls for tankers, but thousands of guillemots, razorbills, and gulls paid the price.

My childhood books shaped my life but even then, there was a certain foreboding. I could never have imagined that the impact we are having today would be so extreme. My dream was to catch a salmon, and the question became, how could I earn enough money to wade the hallowed pools in Scotland? Not, would there be any left to catch, which is the thought that will be entering the mind of any youthful fishing enthusiast today. I never read children's books, instead immersing myself in the science of nature and endless lists of creatures and their habitats. I only experienced the delights of Roald Dahl when reading to my children. I did read The Hobbit, but it was a bit too much fantasy for me – I mostly wanted books that were full of science and facts. Although somewhat stretching the definition of fact and wandering into fiction, it was the animal stories on the edge of fact and fiction, where the descriptions of flora and fauna were true, that piqued my interest: from The Jungle Book, to the Ring of Bright Water, Tarka the Otter, Salar the Salmon, Black Beauty, Wilderness Ways by Paul Annixter and Wild Animals I Have Known by Ernest Thompson Seaton, with his wonderful characters like Lobo the Wolf. There was a certain amount of anthropomorphism without doubt, but their lives were real based on the author's knowledge and their love affairs with nature. I progressed into my teens with the romance of Thomas Hardy, but again there was a rich and accurate vein of natural history weaved in among the challenged characters, as my heart oscillated between a young Julie Christie as Bathsheba Everdene, and could I one day cast a Spey line upon a Scottish river in a named pool with a Thunder and Lightening. What a thought!

It never occurred to me that in my lifetime all those common - and I mean common - birds in Coves Lane - the partridges, lapwings, cuckoos, skylarks, turtle doves, spotted flycatchers and even house sparrows - would simply melt away. So "What's in a year" is also a record of 2023 and what is still here, but there is hope - because conservation is impossible without hope. I feel immensely privileged to have spotted flycatchers nesting every year around our house, but it would be nice to not have to fret each May that they may never return.

There are well-documented statistics about nature and nature loss: for instance, how we have lost 97% of our Wildflower meadows, 200,000 miles of hedgerows, and half of our ancient woodland since World War II. Big data is broadly shared but at a more granular level, do we all really notice fewer birds here and fewer species there? Increasingly as we have become more and more urban in our outlook as a society, a whole generation has little to no understanding of what they are really missing.

You may ask, why does it matter if there are fewer swifts screaming around the village each summer – who cares? Forget the science and their place in the wider ecosystem, I care because they touch my senses each spring and bring joy - I look forward to their return, I marvel at their black shapes, the music to my ears, and their formation flights. They are part of my annual cycle of senses and help make me who I am. Do children now grow up with the sound of the first cuckoo, the first swifts, and turtle doves? Already I am of a generation that failed to hear a corncrake. That saddens me, so I write to share what is still here today and embrace and cherish what is on record. All around us, the world is changing: what was once common is now rare and we become accustomed to a new normal without realising it creeping inexorably upon us. I want to put down a marker on where we are today and prepare for change but equally to share what can be enjoyed in the new norm and how we can and must adapt.

To be a point of reference and provide some structure, I have divided the book into twelve chapters, to represent the seasons' monthly opportunities in fieldsports, nature, gardening, and cooking (apologies upfront if it is a little biased toward fishing!). And, while I am indeed fortunate to be able to travel to some wonderful places, most of the wildlife, cooking, and gardening commentary is based on my life here at home in Wiltshire. In every month of the year, each one of us can experience a flower in bloom, and there will always be some birdsong to lift our spirits and something inspirational to find in the kitchen garden. There is some joy to be found in nature all year round.

There are many questions to be asked as we adapt to a changing environment and clearly we don't yet have all the answers. Like it or not, over centuries we have created a managed landscape, which has, over this time, contributed to the diversity of the creatures around us. If we attempt to rewild, we have to let our managed landscape down gently. The world is full of well-intentioned people who do not necessarily understand the consequences of their actions. Give me a countryman any day, who lives, breathes, and sees the land every day, over a national government body with no real in-depth in-person experience. Equally, we in the country also need to get our acts together in better communication and the rooting out of bad practices. The new buzzword is 'regeneration' and I like to think it incorporates making the most of what the landscape has to offer while recognising we have no choice but to play God sometimes, with six billion of us on the planet.

The gap in our knowledge between town and country - or modern life and nature - widens with each generation, disconnecting us further from the very landscape that supports our wealth and wellbeing. This is not helped by the fact that our urban population has swelled by over 15 million since 1960 and the rural population has shrunk by nearly a million. Nature, as a formal part of a child's education, is missing in action and cannot be mitigated by a few brave and passionate teachers or naturalists. As a Senegalese forester said at a Global conference on conservation way back in 1968:

"In the end, we will only conserve what we love, we will love only what we understand, and we will understand only what we are taught" – Baba Diem 1968

He was spot on, and we did not listen. But there is hope - we finally have approval in England for a new GCSE in Natural History and are in the process of developing a follow-on A Level to provide a pathway to nature-related courses at university*. This is an incredibly exciting development and an important milestone. It is our hope that this GCSE will inspire students to become naturalists and give them the grounding required for environmental studies, not to mention the multitude of green economy jobs that are being created in their droves. We have successful civil engineers, construction experts, and a multitude of good-minded, well-intentioned people, however, without a deep understanding of nature, they are not necessarily making the right decisions. The new GCSE (and A Level) will focus on providing a better all-round understanding of our ecosystems and the wider context, enabling better and more informed decision-making. I would also hope that a more nature-literate society would be a happier society.

The 2019 IPBES Global Assessment Report on Biodiversity and Ecosystems estimated that over $7.1 trillion of global industry value is highly dependent on biodiversity. Yet we are not teaching our children about nature, the very thing that sustains us and drives our economies. How do we address the statistics below if we are uninformed?

95% of the food we eat is grown in topsoil that is being degraded by human activity.
Global Assessment of Soil Pollution 2021, FAO and UNEP

Global wildlife populations have declined by 69% on average since 1970.
Living Planet Report 2022, WWF for Nature

40% of approved medicines in use today are nature-derived, including 70% of all cancer drugs approved since 1980.
World Health Organisation, Cancer Research UK

Approximately 75% of global food crops rely on animal pollination. The annual estimated global economic value of pollination is now up to $735 billion**.
Assessment Report on Pollinators, Pollination & Food Production 2016, IPBES.

I hope and intend to push for a renewed understanding and passion for nature with

*Unfortunately this has now been put on hold by the new Government
**Adjusted for inflation to 2023

all interested bodies, both Government and Industry, by using the GCSE For Natural History as a pathway to a commercial qualification recognised in business. With more and more companies employing a Head of Sustainability and the agri-food industry seeking sustainability as the holy grail, incremental new job opportunities are emerging in Biodiversity Net Gain (BNG) and Finance and Accounting. These developing nature markets can help join the dots between the GCSE and business, reconnecting us back to nature.

SOME OPEN QUESTIONS FOR DEBATE

When it comes to the countryside, I have realised that opinions are often too polarised to make progress. When I was writing this book, it was a relief to be able to ask questions and quote opinions, after years of having to be professionally 'on message' and aligned to a particular organisation's official stance.

Below, I call out some open questions for which I don't have all the answers, but I have aimed to give some wider context as these themes pop up later in the book.

SHOOTING AND BADGERS

Is there a limit that should be prescribed for a maximum density of game birds put down per acre? I am not sure large game bags are always defensible and I think we should be managing our homework a bit better. Overall, shooting is a good thing preserving our countryside flora and fauna in many ways, but are we killing the golden goose by exceeding the holding capacity of the land with an excessive number of both birds and shooting days? Business thrives on scale and there lies the problem, scale and the countryside, culture, and conservation are not necessarily good bedfellows.

Badger numbers have rocketed at the same time our hedgehogs, bees and ground-nesting birds have declined. Is it a coincidence that there is now a plentiful supply of badger food - either roadkill or unpicked shot game - precisely at the time when they would normally be feeling the pangs of winter? One of the benefits that badger protection brought was the banning of badger baiting with dogs, and quite rightly so - put in the dustbin along with cockfighting and bear baiting. But we never imagined that there would now be an estimated 500,000 badgers, having doubled in number in England and Wales since the 1980's; nor considered what impact this would have on our local ecosystems.

Nowadays, I prefer the simplicity and honesty of stalking, either I take the shot, or I don't - either way, the deer's life has been natural, and I am just taking a surplus for the table with no impact on the wider environment.

OUR GARDENS:
BIRD TABLES

Many of us are self-proclaimed 'nature lovers', and as agriculture becomes more industrialised, our gardens become increasingly important bastions of biodiversity. But does the estimated £200 million - £300 million we spend each year on bird food make our bird tables super-spreaders of disease for our beloved wildlife? Do these bird tables make our wild birds more susceptible to constant attack and disturbance from our much-loved dogs and cats? And with a dog and cat population of over 24 million, can we legitimately say we are nature lovers, bearing in mind the subsequent increased (and so often overlooked) pressure that our pets put on the environment – whether they physically kill animals or not, their very presence causes disturbances to nature.

And are we unnaturally inflating predator numbers beyond the normal prey and predator balance by providing supplementary food on bird tables? And yes, squirrels and rats are predators, along with many domestic cats and dogs.

The problem here once predators' lives are out of kilter with their prey and that very same prey is already under pressure from agricultural practices and urbanisation, predation becomes the final tipping point. Frustratingly predator control is often a taboo subject, with those opposing it often claiming that it is 'playing god'. However, if we have created the zoo, so to speak, then we must be the zookeeper.

GARDENING

What is gardening today and why do we still crave a green-striped lawn? Incredibly, it goes back to a rich man's fashion in the late nineteenth century. Do we want to replicate the sterile, single monoculture fields we see today, constantly harried, and sprayed by beasts on wheels in our gardens? And yet we say we love daisies and clover and bees and buttercups? The trend is moving towards wilder and less manicured gardens with the debate now broadening, recognising the importance of variation - so a mix of some wild, some less wild and some manicured gardens is a step forward and avoids the creation of polarised arguments, which is so detrimental to any cause. In any case, even wild patches require a degree of nurturing.

FROM GARDEN TO KITCHEN

I find growing my own fruit and vegetables, while following the seasons hugely inspiring, interspersed with fish or game I have caught or shot. I am naturally omnivorous and have no quarrel with a vegan. However, I cannot get my head around the argument that all beef is bad when we live in a grass-growing country, and yet it is acceptable to fly water-hungry avocados halfway around the world? If we want to eat sustainably, surely, we should prioritise how we look after our soil, water, and food miles. Start local, have good husbandry and clear provenance. In this vein, I can only defend shooting a gamebird if someone ultimately eats it, otherwise, I have lost the argument.

We are blessed to live in this world and we who practice fieldsports, cook, garden, and enjoy nature - both flora and fauna - have a unique view of how our world is changing. For the sake of this planet, to which we are bound, we must open our eyes and really look. Your lives will be even more blessed. Immerse yourself in the present moment - we are all deeply connected to this place.

As I lament the drastic changes I have seen even in my lifetime, hope can be found because we know so much more today and have so many emerging technologies to help protect what we see, as long as we keep looking!

Mrs. Owl keeps watch from an old Scots pine tree.

APRIL

MRS PHEASANT AND MRS OWL

OUR NEW NEIGHBOURS: THE TAWNY OWLS

"Good night girls," I say in hushed tones as I shut up the chickens for the night. There is a chill in the air, close to frost, and as I stepped out in just a shirt, I turned to hurry back to the warmth of the house.

A long, drawn-out "tisss" followed by a pause, then another "tisss," stops me dead in my tracks. The sound cuts through the stillness, sharp and unexpected, making me hesitate and listen intently to the night air. Straining your ears is an entirely different experience from focusing your eyes. You instinctively reduce any unnecessary movement—slowing your breathing and trying to remain perfectly still, immersing yourself in your surroundings, avoiding any slight rustle of your clothes. A living radar station.

There it is again "tisss". I walk slowly toward a small clump of old Scots pine trees, my footsteps cautious and deliberate, still not quite believing what I'm hearing. Each step heightens the sense of anticipation, my excitement growing with every gentle crunch beneath my feet. It's as if the night air holds its breath, amplifying each sound,

Right there at my feet, nestled into the base of one of the Pines, was a little ball of fluff about the size of two tennis balls—a baby tawny owl. It clicked its bills together instinctively, a tiny but fierce warning. This sound was immediately answered by a much louder, intimidating, and threatening snapping noise above my head. Peering up into the fading light I could just make out the silhouette of an adult tawny owl, clearly agitated by my presence near its young.

Caught in this raw, primal tension, my mind flickered between the urge to inspect the little fluffball at my feet and how to mitigate the potential danger from above. I couldn't help but think of Eric Hosking, the legendary bird photographer who lost an eye after being struck by a tawny owl while attempting to photograph its nest. It was a reminder of the fierceness with which these owls guard their young, and the fine line between curiosity and respect when you find yourself stepping into another creature's world. This is night—its domain, not mine. I am the outsider.

In addition, I was struggling to grasp how early in the year it was, a little unseasonal for an owl to have already left the nest —only April 7th—and the incredible realisation that this entire encounter was unfolding right in my garden. It felt surreal, as though nature had crossed an invisible boundary.

The truth was that it was I who had crossed that invisible boundary, by carefully creating a patchwork of habitats. The mosaic of varied grass species and lengths attracted the mice and voles, which together with minimal disturbance drew in the owls. I welcomed nature back into this space. It's often that missing element—quiet—that allows wildlife to flourish.

I withdrew quietly and summed up the situation. *"Always leave young birds where you find them"* I kept saying to myself, *"and anyway tawny owl chicks often leave their nest before they can fly"*. But this fledgling was too young to be on the ground, so I decided to place it out of harm's way up in a tree, but by the morning it was back down on the ground again.

The baby owl in her crib.

APRIL

I decided it was time to play God and placed a wicker basket five feet off the ground, sheltered against a fir tree with the little owl safely inside. The next morning it was alert with a dead wood mouse next to it – a great result. The adult owls had found their owlet and ensured it had provisions. These owls make a *"tissssss"* noise almost continuously at night so the adults can locate their fledglings throughout the hours of darkness until they are well fed. Amazingly, at twilight the following evening I watched a parent call *"kewick"* and the little owl in the wicker basket responded *"tissssss"*. At that moment, the male appeared silhouetted against the fading evening sky and passed a dead mouse to the mother, who flew down to the wicker basket.

And here we are three weeks later, all doing well and see how it's grown. I am conscious these beautiful birds are only here due to our careful garden management and the lack of disturbance.

For the next few weeks, the young owl was often quite exposed sitting in an elder bush, but its mother was never far away, and no self-respecting crow or magpie dared to show it any attention. It could not fly yet but I often found it had moved trees, so it must have walked – or should I say hopped - across the grass and climbed vertically up the tree trunks. Amazing!

Out of the crib and growing strong in one of our Rowan trees

PHEASANTS IN THE KITCHEN GARDEN

It has been a good year for observing nests and on 16th April I found a hen pheasant in the kitchen garden next to four eggs. There is nothing too unusual in this except for the fact that once again it was a little early in the year.

The eggs were somewhat hidden among some bluebells, but still quite exposed as she was under the mulberry tree which is one of the very last to come into leaf each year. However, by the 21st of the month, there were seven eggs, which over the coming days she added to at the rate of not quite one per day. It was not until the 8th of May that she settled down to incubate 15 beautiful shiny olive-brown eggs about half the size of a chicken's egg but more rounded.

The nest was no more than a slight scrape in the ground with a couple of her feathers and some squashed bluebell stems. I grew increasingly invested in this hen (affectionately nicknamed Mrs P) and her soon-to-be brood, as luck would have it, she was visible from my office window. On top of this, every day at around 3pm she would do a little Jemima Puddle Duck foray over the kitchen wall and into the fields for no doubt a wash, feed, and drink.

APRIL

The April we had in 2023 was quite cold and frosty, which a bird can cope with, but May was bleak with unusually cold and extremely wet weather. With torrential showers and hail, I regularly thought about her eggs laid bare on the ground with no soft eiderdown to keep them warm. Despite the adverse weather, Mrs P. would sit there patiently, often soaked like a washed-up seabird on the shoreline. Days went by and we wondered whether the eggs would survive the unseasonably cold weather.

Twenty-one wet, miserable days passed - during which I planted out the peas, sweetcorn, and courgettes. The potatoes came up and I sowed carrots and beetroot, all the while giving Mrs P. the widest berth possible, although she did not seem apprehensive about human disturbance. I had to chase away a carrion crow one morning and I always kept the gates to the Kitchen Garden shut to keep out any potential predators. That said, it proved to be a relatively safe place.

On the Sunday 23rd of May, she had a little twinkle in her eye - and yes, I am guilty of being anthropomorphic and a little melodramatic here, but I had become attached to this pheasant, as well as the owl at the bottom of the garden. I noticed that she kept tilting her head to one side and realised she was most likely hearing the chicks in their shells. Sure, enough, on the Monday, ten little bumblebees of birds appeared from under her tail and out of her wings, as well as the surrounding vegetation.

That first day Mrs P. stayed very close to the nest keeping the chicks dry in the showers and still brooding the remaining four eggs. Usually, pheasants move away quickly from their nests once the chicks are hatched to avoid predators, but while she was able to fly up onto the kitchen wall, her chicks could not yet follow.

I kept the gates to the garden shut for the time being as I figured they were safer in the Kitchen Garden for their first day in the big wide world.

And so began an assault on my mangetout; and then Mrs P. ended up brooding the chicks on that first night between the purple sprouting broccoli, which took an obvious bashing, but the row of parsley next to it was left undisturbed – clearly it was not to her taste.

By 6:30am on the Wednesday she had not moved, and it was only at 8:30am when the sun came out that the first chick started to venture out. They all took a walk through the strawberries, before settling down among the garlic stems where there is a bit more cover – they have been growing since November. As the day continued, the chicks grew bolder and more certain on their feet. Their mother would dig with her beak and call them to look for invertebrates in the broken soil while she stuck to a vegetarian diet of broad beans, peas and mangetout. I will also have to re-sow the carrots and beetroot. I do hope at least the chicks are finding all those slugs that love to devour my seedlings!

Finally, early on the Thursday morning, I found the little family awake, walking around and pecking the ground in search of food. It was amazing to see the increase in confidence levels after only 24 hours. I decided to finally open the gates from the Kitchen Garden to beyond. However, it was not until 3pm when the little pheasant family finally plucked up the courage to strike out to the big, wide, treacherous world – I felt a little sad.

It was not until May 8th that Mrs P. finally sat down to incubate her 15 eggs.

APRIL

The month of April is the first month of the year when you can finally hear the summer migrants joining the dawn chorus. Often, it is the two-tone chiff chaff and serenading willow warblers who first make their presence known. Willow warblers are now far less common in the south of England – they seem to have moved further north – and I miss their constant low, sweet notes as they hunt out the early nectar in the pussy willow. Their subtle green and yellow feathers merging into the bush's grand flowery show.

These days, you are more likely to hear the melody of the blackcap - a sheer flute-like song punched out from a smart grey bird with a sooty black hat, roughly the size of a robin. If you leave a little area of the garden wild, ideally with some blackberries and long nettles, with a little bit of luck, you will hopefully have their sweet notes serenading you throughout the spring. I have noticed that the blackcaps in my garden often sing among the wild cherry trees in bloom, contrasting against the white blossom with their white, grey and black feathers. They are constantly fidgeting among the blossoms: another branch here, another branch there.

Towards the middle of the month, you may still hear your first cuckoo. Their populations have declined, particularly in the south of England, where they seem to increasingly favour watery locations with reed warblers now an important host for their eggs. Further north they appear to be faring better and are heard on the moors, with meadow pipits being their preferred hosts.

Interestingly, recent research and tagging suggest that a southerly migration route that takes cuckoos down to their winter quarters in Africa via Italy is a more beneficial journey with comparatively fewer deaths than the alternative western route via Spain and Gibraltar.

PREJUDICES & IRONIES: GROUSE SHOOTING, PEATLANDS & CONSERVATION

I often witness a frustrating irony among people who never bother to look at the provenance of their supermarket fish or meat yet become incredibly righteous and antagonistic over people shooting animals to eat.

On top of this, there is an additional irony in their agitation over the control of certain predators, such as foxes, stoats, and corvids[1], when no consideration is made to how many wild animals die to produce their mass-produced supermarket meat – do they have any idea how many wild fish and farmed salmon die, for instance, to produce one kilo of farmed salmon or any idea about harmful practices involved in the majority of chicken production.

[1] *Corvids include carrion and hooded crows, magpies, and jackdaws.*

I think the shooting industry needs to learn to communicate and educate better in the modern world and root out any bad practices where they exist. Especially when you consider that wild game is part of the food chain - just like any other meat that we consume in the UK – and so there must be standards upheld to ensure best practices concerning both animal welfare and the quality of the meat given its destination is for human consumption. However, some of the most fractious arguments for and against game shooting are centred on grouse shooting which I find the greatest irony of all, as it is one of the most sustainable field sports we have in the UK today. This is, after all, a wild creature.

Naturally, people focus on August 12th – the start of the grouse shooting season - and the physical act of shooting red grouse, but this is the culmination of a year's work and is entirely dependent on the success, or otherwise, of the breeding season. If chick survival is low there may be no surplus birds to harvest, and consequently, no shooting will take place that year.

This current 'spring season' is the other, and often unseen, side of grouse shooting. A moor alive with the sounds and wings of waders and summer migrants. We often focus too much on the actual killing of a bird (in this case, the act of shooting) and in doing so, fail to consider the ecosystem in the round.

A DAY OF LEARNING IN THE NORTH PENNINES

The provenance of one 'good news' story lies high up on the Pennines. I have travelled to the north of England, and it is here I experience a snapshot of the past – or as I see it, an island of hope for our breeding lapwing, snipe, curlew, redshank, golden plover, and black grouse. All these birds are critically dependent upon managing the moor for red grouse shooting; and whilst these moors continue to be carefully managed, so too do these birds continue to thrive.

It is now 4:30am and I am hidden away in my hide; I am blind to much around me in the slowly creeping, dim light of dawn. My body is aware only of the cold; and my ears, only the continuous hissing and bubbling from the moor outside. I am crouching here in this hide to witness a 'lek' (derived from the Swedish word for 'play') - when the male black grouse fight, perform, and dance to attract a greyhen (a female black grouse) with which they can then mate.

For those who are new to the world of grouse shooting and/or birdlife, it is worth mentioning that black grouse differ from their better known relative, the red grouse. Considered endangered by the British Trust for Ornothology (BTO), black grouse are now reduced to living on the edge of moorlands, (or should I say have been pushed to the edge of moorlands, considering they used to be present in every English county) with an approximate population of 1500 adult male birds in England, according to the latest BTO census – of which the majority live here in the north of England. As a result of their low numbers, these birds are not targeted for shooting like the red grouse but instead are a wonderful beneficiary of the managed moor.

Back to the cold and dark of my hide and I can hear the male black grouse practicing their moves and sounds (a mix of guttural hisses and 'rookooing' created by air sacs in their necks) with their sparring partners as they wait for the arrival of the greyhens. Rising slowly in the sky, the sun reveals their shapes: pure white undertail feathers and dark blue plumage, contrasted against their scarlet eye combs. They are incredibly striking birds.

A male grouse: one of the UK's most wonderful birds, yet its survival is threatened. Photo by Ian Harrison

As you first scan the land in front of you from the cover of your hide, your eyes can only just detect the white under tail feathers of the birds, reminiscent of white fluffy rabbit tails - but suddenly, the black cocks spring into the air with a flurry of wings, alerting you to their presence. The cocks have spotted a greyhen from afar who lands quickly in their midst, before keeping her head down among the ruckus of brawling, skirmishing males, all now trying to catch her attention. The young first-years attempt to join in from the outside of the lek only to be aggressively driven out by the older, more dominant males. In all the melee, the occasional greyhen is mated, and the ritual is accomplished. It is a privilege to witness this seasonal event and encouraging to see their numbers at least here holding steady.

Among British birds, only the even rarer capercaillie perform these leks on traditional open spaces year after year. Photo by Ian Harrison

APRIL

Leaving the black grouse in peace at 7 am, the now waking world is full of waders busying themselves with their morning activities. The lapwings are already shepherding their little balls of fluff through the grass and making their looping alarm flights whenever danger threatens. Although apart from the odd lesser black gull and carrion crow, here at least, the weather is their greatest danger. They have found peace to rear their young.

A snipe flies fast and high. He is on a mission as he arcs across the sky before tumbling down with a whirring sound called 'drumming' (a noise that is reminiscent of a waving lightsabre in Star Wars) as the outer tail feathers vibrate. The faster the snipe plummets, the louder the sound and the more attractive he appears to his suitors below, hidden somewhere in the rushes.

A redshank stands on the stone wall lining the track, his reddish-orange legs starkly contrasting against the grey background. And, true to his role as 'sentinel of the marshes', he rises with a startled *"tewk tewk"* before riding the wind with a longer drawn out *"yew-yu-yu-teu- teu toooooooo"*.

But above all, it is the curlews that are the star attraction, wings outstretched with their mournful *"pree pree"* before a full serenade builds and builds *"prep prep prep preeeeeee"*, their song echoing across the hills. The sound transports me back to the countryside books of my childhood by 'BB' – stories complete with stone walls, hedgehogs, and curlews, teaching us all about the seasons' circle of life.

Behind the pseudonym *'BB'* was Denys James Watkins- Pritchard MBE, a British naturalist who won the 1942 Carnegie Medal for British children's books. Through his written words – illustrated only with simple black and white wood cuttings - he inspired a generation of nature lovers. The name 'BB' was based on the size of lead shot he used to shoot geese, which wonderfully illuminates the acceptance of both life and death, an understanding that we have lost today in a well of sentimentality.

Back to the present, and the birds here that I can see today in late April are a direct result of this acceptance that life and death go hand-in-hand, as well as a more sympathetic management approach to the landscape. The hard truth, is these birds are on an island of sorts - the area around their unique moorlands washed away by agriculture and development. On top of this, a frustrating deafness, a sad denial, and an unwillingness to accept that grouse shooting, with the funds that it provides, sustains this wildlife utopia. These waders so alive - thriving here on these moors - have all but been lost or greatly diminished from lowland England plus the higher Dartmoor, Exmoor, the Lake District, and most of Wales. It begs the question, what is different here?

A UNIQUE HABIT

These moorlands (or peatlands as they're often referred to) in the northern Pennines are classed as blanket bog[2] and are vitally important both for our wildlife but, also for climate regulation.

With historically little agricultural value, most of the public had no idea how important these bogs were until recently. Thankfully, though, we now understand the crucial role they play in climate regulation, and increasingly more is being done to ensure these unique landscapes are protected.

That said, while many people across the UK understand the importance of forests when trapping carbon, few realise how peatlands are among the most carbon-rich ecosystems on Earth. In fact, the world's peatlands trap and store twice as much carbon as the world's forests, having been formed approximately 12,000 years ago during the Ice Age.

For this reason, we must leave peatlands undisturbed: not only do they continue to store billions of tonnes of carbon, but they also have a net cooling effect on climate and slow the flow of water-cleaning it naturally and reducing flood risk. Equally, these peatlands support a unique and diverse habitat, increasing bio-diversity.

The red grouse is our only endemic bird in the British Isles, surviving primarily on a diet of heather. Until now, this unique environment has been specially preserved, funded by the red grouse shooting industry.

THREATS

The move to offset carbon by planting trees, although laudable, is disastrous if the trees are the wrong species, and or, planted in the wrong place. From the point of carbon sequestration, no trees should ever be planted on peat. It would be completely counterproductive. Instead, more trees should be planted in the 'right' places and especially along the 'becks', 'burns', 'gills', and streams to shade the water, cooling the temperature to provide some mitigation from global warming.

Perhaps the bigger threat is political and cultural, with increasing pressure on how these moors are managed.

Traditionally, heather is managed by undertaking a process known as 'muirburn', where localised areas are burnt to create a mosaic of different age, growth, and habitat

---

[2] *A blanket bog or blanket mire, also known as featherbed bog, is an area of peatland, forming where there is a climate of high rainfall and a low level of evapotranspiration, allowing peat to develop not only in wet hollows but over large expanses of undulating ground.*

edges, that favours grouse. Professionally managed, the argument is that the heat does not penetrate the wet soil and so there is no damage to the underlying peat. In recent hot and dry summers, some unmanaged moors have gone up in smoke, where a build-up of old grass and heather has provided the perfect accident waiting to happen.

Some management of the moors is required. It seems the direction of travel is probably more mechanical cutting rather than burning, although this is challenging in the remote hills. It may be a case of 'perception matters more than reality', as the sight of moors cloaked in billowing smoke is not a good look from a PR perspective - particularly in an era of social media where a photo does not necessarily provide the full picture.

On the moor where I took these pictures, sheep are grazed at low densities, and importantly, brought off the hill during the winter to allow the grass to recover, which ensures optimum nesting conditions for the waders. High densities of sheep across the moors all year can lead to a degradation of this unique habitat, with a short uniform grass sward leaving nests more susceptible to predation.

This brings me to the other contentious issue of predator control. Like it or not, we live in a managed landscape, and it has been proven time and again that if no predator control takes place, then these wading birds and grouse decline rapidly.

In fact, managing predation is also vital to ground-nesting birds of prey, such as harriers, merlin, and short-eared owls, who, in the past, have unfortunately experienced instances of illegal persecution on grouse moors in a bid to protect the grouse. I do believe this is minimal, but it must stop. It would be the saddest irony if, because of a few bad actors practising illegal predator control, all predator management was banned - with not only the grouse and waders suffering but ultimately the ground nesting birds of prey themselves.

Finally, the sporting world can also help itself with better communication - and perhaps by adopting a more modern lexicon. The management of these moors is still dependent on an increasingly aged population of proud gamekeepers who feel their lives and livelihoods are continually under threat. Maybe a change in title to 'wildlife ranger' would attract younger recruits and better define the true role and value that these people perform out on the hill every day.

*Curlew in the North Pennines with its charismatic long curved bill.
Photo by Emily-Graham-media*

APRIL

MY PERSONAL VERY SIMPLIFIED TAKEAWAY IS SUMMARISED BELOW:

MOORLAND PEAT: Re-wet the peat by taking out the 'grips' ('grips' is a term for previously encouraged drains on the moors) and actively discourage tree planting on peat.

OBJECTIVE: Improvements in both water and carbon retention.

REDUCE DEER AND SHEEP DENSITIES: This includes bringing sheep off the hill in winter.

OBJECTIVE: Improved revival of trees and a mixed grass sward in the Spring, beneficial to nesting waders.

PLANT THE RIGHT TREES IN THE RIGHT PLACES: Mixed native pine and broadleaf along the river alluvial soils and tributaries.

OBJECTIVE: Shading and temperature reduction of the rivers and streams.

PREDATOR CONTROL: Maintain an active control of foxes, corvids, stoats and weasels.

OBJECTIVE: Increase nesting success of ground-nesting birds, including grouse, waders, songbirds, and birds of prey.

HEATHER MANAGEMENT: Where practical replace heather burning with mechanical cutting of the heather.

OBJECTIVE: Maybe more challenging than 'muirburn' and the science inconclusive, but the risks of burning are high and politically smoke-filled moors is damaging PR.

NON GROUSE MOOR HABITAT

We do not have the luxury of providing grouse moor sanctuaries everywhere, but for ground nesting birds to be successful, all NGOs now accept the 'three-legged stool' principle of habitat, food and legal predator control. If we agree that grouse moors will always be utopia for waders, what is an acceptable level of breeding success outside of these moorlands?

There is some criticism that in public spaces such as National Trust land and RSPB reserves, although spending is high, the success rate for breeding waders is very low. However, the acceptance of some predator control, habitat management, and fencing of nest sites is beginning to pay dividends on RSPB reserves, such as West Sedgemoor in Somerset and Glenwherry in Northern Ireland.

What is desperately needed, are options in ELM (Environmental Land Management) for both predator control and grass management, agreed by DEFRA and the Government, which ultimately filters down to rural payment schemes for farmers. With the best will in the world, habitat on its own is not the answer.

LEAN INTO THE COLD TO CATCH YOUR FISH

After two or three casts; ice begets ice, and you have to break the build up from your rod rings

APRIL

There is one place I know where the sea trout will take your fly this early in the year. And surprisingly it is further north, way north, in a small two-kilometre spring-fed stream called Tungaelinkur in Southern Iceland.

As you drive southeast from Keflavik on an early April morning, only the lengthening days signal a change in the seasons. The grass is still a light flattened ochre, dry and dusty in the -6C wind chill coming straight down from the Arctic. The first whooper swans seek out the plants submerged in water to refuel after their long migration, while a few golden plover and oystercatchers wait sentinel-like, preserving energy until the watery sun thaws out their breakfast. They look out of place in their stark, smart spring plumage, incongruous with the drabness of last season's dried and seemingly dead grass.

However, it is the geese that really catch your eyes and your ears. A constant honking of Greenland white-fronted geese signals their arrival from the UK and Ireland to this one relatively sheltered spot on the island. If you are captivated by bird migration, these enchanted birds are even more amazing, as they are only stopping here briefly before journeying on to Greenland. Upon their departure, the white-fronted geese are then replaced by our own wintering greylag geese who spread-out all-over Iceland.

The birds are the only signal that the land is waking from its seasonal slumber, all other signs point towards winter, and in the distance the vastness of the Vatnajokull ice cap stares down, brooding, reminding you of your vulnerability. And so it is - I must admit, with a little scepticism – that I find myself checking out the Tungaelinkur. With the winter cold keeping the water locked in the catchment and frosted margins there is but a minimal flow; so, I am astounded when a fish swirls, then another and another.

Everything about this river is different from the UK. The sea trout arrive in September, fresh from feasting on nutritious sand eels that thrive in the glacial-fed waters off the southwest coast. The sea trout come and go, back to sea, and back to spawn many times, and grow big as a result. Iceland has never allowed sand eels to be commercially fished, preserving them as the mainstay of the ocean food chain that supports so many species - both fish and birds.

Upon entering the river, around half the sea trout will spawn then and there, while the remaining fish stay over winter, locked under its ice in preparation for spawning the following September. It is these over-wintered fish we are now searching for during a short one-month 'spring' season before the river closes again until September. Clean water, plenty of food, little fishing pressure, no farming or agriculture along the banks and 100% catch and release. When I say the river is spring-fed, the water is filtered through the ever-present lava, producing the clarity of a chalk stream before it enters the larger Glacial sourced river Skafta.

Within an hour, dogmatism has replaced scepticism, and we understand what a gem of a river this is.

One of the joys of fishing is coming close to these beauties – a wonderful cock sea trout.

APRIL

SUMMER MIGRANTS

A wet January and February have given added depth and flow to the river Kennet this year, smothering the water with long waving grass and crowfoot like the Okavango. West Overton in Wiltshire is a crucial stopover place for some summer migrants to rest and feed. I imagine the extra energy the birds require to fly up from the Pewsey Vale to cross the watershed from the catchment of the river Avon. Down they then fly to the river Kennet – a perfect place to pause, rest, and feed.

During a cold spring, the air is full of hundreds of small birds making the most of what seems to be the only place to hunt. And today on 20th April there are over 100 swallows in a small area of roughly an acre, wheeling over the waters and twittering above on the telephone wires. They are accompanied by about 20 house martins and two or three sand martins.

A flash of yellow catches my eye and yes, there is a yellow wagtail and as I get my eye in, I can point out two more on a wooden fence upstream. A birder's dream but it is only me here to witness this moment and I quietly rejoice. These tiny birds are fresh from sub-Saharan Africa all revelling in our tiny corner of the globe. This bridge and this little meadow are making a difference to these creatures on their long migration.

In the water, a mallard is feeding close to a pied wagtail bobbing on the floating crowfoot with its dainty yellow and white flowers, together with a couple of moorhens. In this moment I am so very present, rejoicing at the incredible flora and fauna around me. It is a moment in nature when I am able to really step back and soak it all in - a fleeting reminder to me of the possible. The western sky was lit up with the last rays of the sun and everything seemed alive and so full of potential. Every few minutes a car would cross the small bridge and I watched the faces of the drivers, completely intent on their destination and task at hand, everything else blocked out with no time to see the beauty and the wonder, the masterpiece they were unwittingly driving through - their expressions without expression.

The next day, I returned but not a bird was to be seen; the pulse of migratory birds had continued north on its journey. Blink and you miss it. The single significance to this beautiful location and the role it plays in the birds' migration is water. Although, by summer like the Okavango its lifeblood will be sucked dry, leaving a scorched mud scattered with sheep shit. Water is so precious. Yet we continue to use more and more of it as we suck the aquifers dry with everything we do in this modern world, not to mention the cities we build.

Under the lime trees in our front garden is the perfect habitat for these wood anemones.

By April, the brook ranunculus are flowering at the head of the Kennet, long before its bigger cousin downstream which requires water all year round.

SAVE THE SPRING: THE ATLANTIC SALMON TRUST & THE RIVER DEE TRUST

When I look back at old photographs of salmon that we caught on the river Dee only ten years ago, many were noticeably fat and stocky with extra-sized pectoral fins and tails. We now know via genetic studies these are a subpopulation unique to the Upper River, designed to better cope with their environment over millenniums. These fish typically entered the river earlier in the year, providing sport from February to May, and running up the snow melt to find some dark hole to wait out the summer for spawning in late autumn and early winter in the upper tributaries. It is these very fish that are now in great danger of extinction; so great in fact, that radical solutions and interventions are now being put in place, including a twenty-year two-part joint program run in partnership between The Atlantic Salmon Trust, the River Dee Trust, and the Dee District Salmon Fishery Board to save these magnificent creatures.

I feel extremely privileged that, having witnessed the decline of salmon numbers in the River Dee over the last thirty years, I now get to play a small role in trying to save them. When you live somewhere, fish each year, or just go for a walk every day in the same place for thirty years you notice changes. The Dee has also lost many of its eels, pearl mussels, and sea trout over this time and, maybe consequently, I am seeing fewer otters.

Salmon play a vital role in these ecosystems, acting in the River Dee as a keystone species and helping to indicate the health of the river.

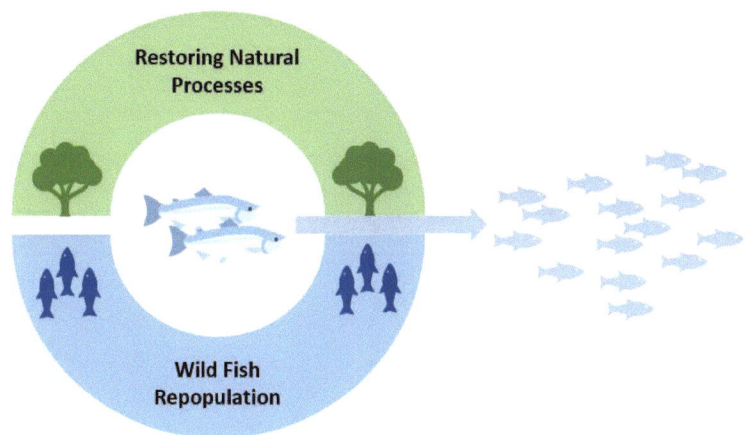

Diagram reproduced by kind permission of Atlantic Salmon Trust

The initial phase of this twenty-year two-part strategy prioritises restoring natural processes within the river and its surrounding landscape. This involves key actions aimed at enhancing the health and resilience of the river ecosystem:

1. TEMPERATURE REDUCTION: Salmon growth reduces as water temperatures rise and by 22C they stop growing altogether. This is not good when we know we require fit smolts to survive their epic migration. Efforts are being made to lower water temperatures within the river, involving the planting of shading vegetation, restoring riparian buffers, and reducing runoff from nearby sources of heat.

2. FLOW CONTROL: The smolts also require the right amount of flow to travel downstream and help them evade predators while the adult salmon require sufficient flows to migrate back upstream and to help oxygenate their eggs. Strategies are being implemented to manage the flow of water from the surrounding land into the river to minimise erosion, sedimentation, and nutrient runoff. Controlling the flow of water can also help mitigate the impacts of flooding, which has recently led to many redds being washed away.

3. HABITAT ENHANCEMENT: Efforts are made to provide more suitable habitats for salmon. This includes creating the right forms of shelter, protection from predators, diverse in-stream habitat features, and ample food sources such as invertebrates. The focus here will be on woodland restoration, peatland and wetland restoration, and re-naturalised buffer strips.

These two photos, taken on the Upper River Dee, show the difference that habitat restoration can make, with great progress currently underway. Photo credit: River Dee Trust.

However, it is recognised that habitat restoration alone cannot save the Dee spring fish - nature needs a helping hand. This second part of the programme focuses on wild fish repopulation involving a method I believe is new to the UK. Desperate times require desperate measures.

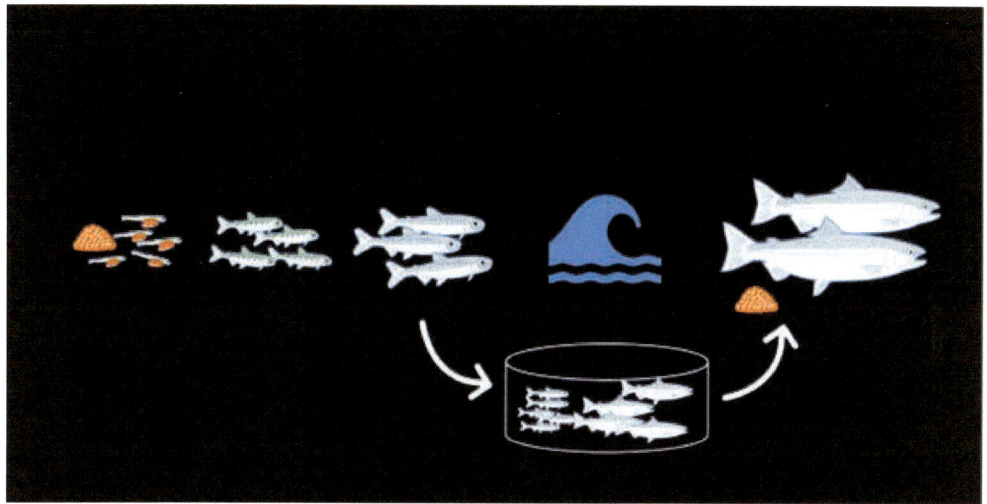

Diagram kind permission of the Atlantic Salmon Trust.

"Smolt to Adult Supplementation," which typically works as follows:

> 1. **SMOLT CAPTURE:** Juvenile salmon, known as smolts or presmolts, are captured from the river before they embark on their spring migration to the sea. These smolts are typically identified and collected using methods such as trapping, and netting.
>
> 2. **CAPTIVE REARING:** The captured smolts are then transported to specialised facilities, where they are housed in controlled tanks with optimal nutrition and appropriate care to support their growth and development.
>
> 3. **GROWTH TO ADULTHOOD:** Over a period, the smolts undergo captive rearing until they reach adulthood and reproductive maturity. This is like 'head starting' in wild birds of conservation concern, which are raised in aviaries until release at adulthood.
>
> 4. **RELEASE AS ADULTS:** Once the smolts have matured into adult salmon, they are released back into the river from which they were originally captured.
>
> 5. **SPAWNING:** By releasing artificially reared adults back into the wild, the

goal is to supplement the existing wild salmon population and enhance its reproductive potential, as left to nature, many of these smolts would have been predated before they were able to spawn.

Wild fish repopulation places an emphasis on maximising wild spawning behaviour and wild hatched juvenile salmon. These two approaches have been chosen as they minimise the negative impacts of domestication associated with traditional hatchery practices and help to preserve the genetic make-up of the River Dee's salmon population, especially within its subpopulations. This is key to enable these salmon to adapt to future environmental change.

The partnership is presently working with the University of Stirling, the Scottish Government, its agencies, and others to develop and refine the wild fish repopulation methods, ensuring that detailed genetic monitoring will enable success to be recorded. After thirty years it is a positive to finally see action, it is just a shame that we previously wasted time blaming everything on troubles at sea.

Mechanical sluice on the river and it still works!

MAY
CONSERVATION IN PRACTICE

May is one of the most magnificent months of the year when it comes to witnessing Mother Nature in all her glory. As her year's work reaches its dramatic climax: from the majestic and flamboyant garden peonies to the subtle, wild, red campion, and across all the lands the beautiful, upright candlesticks of the horse chestnut trees. The show has well and truly started.

The light and water in the river collaborate, keeping the trout hidden from sight. The stream is full to the bank, its surface reflecting and refracting the surroundings, the current is too fast to sight any fish as I make my way down to the old pump house. The conditions make a big change from last September when the same quick scan would allow me to count a dozen fish, each holding station in the few remaining streamy runs - the difference between spring and autumn fishing.

We all love sight fishing, when the perfect cast is lightly dropped a foot above the waiting mouths of the fish; but in May it is a case of judging slight movements in the water, a mere dimple or, if you are lucky, a nominal splash quickly healed by the downstream passing. The constant casts upstream, adjusted for the flow and the constant drag on the line, make for hard work. But, on the positive side, the small window for interception by the fish can give you the edge. With little time to check out the fly and its binding hook, the take by the trout can be more confident and aggressive.

Having spent many an hour on a riverbank, I have noticed that the birds sing more in the daytime when it's cloudy for some reason. And today, not seeing any fish, my ears tune into a newly arrived grasshopper warbler. His reeling high-pitched notes come from deep within the longer sedges that the grazing bullocks cannot reach, while a chiffchaff "chiffchaffs" from the nettles.

Here in this very special corner of Wiltshire you can still frequently hear and see cuckoos, including the strange gurgling song of the females. It is an unusual song but so very apt along the chalk stream of the river Kennet, which some have described as a bubbling chuckle. You will be lucky to hear only a few times in your life, compared to the thousands of times you might hear the classic "cu coo" as the male performs his 'peacocking' ritual. On her part, the female only sings this short bubbling duet with her new groom, who changes his drawn-out "cu coo" to a more hurried "cu cu coo" in his excitement at the prospect of his new mate. Next time you hear a cuckoo listen out for something a little out of the ordinary.

'Mate' may be too strong a characterisation, as the male cuckoo does his best to attract as many females as possible in a short window before he flies back to Africa, sometimes as early as mid-June. Not long after in July, the females follow their male counterparts south once they have laid a single egg in a number of songbird's nest to be raised by their new adoptive parents; here in Wiltshire, cuckoos predominantly lay their eggs in reed warbler nests. It is for this reason that small songbirds dislike cuckoos so much and raise alarm calls in their presence. Another cuckoo calls, and yet again, nature manages to humble me. Opening a small window into their lives.

Still searching for a rising fish, a pied wagtail catches my eye flitting along the bank, launching itself at the emerging mayflies flying erratically from one life to their next, skywards from their watery beginnings. Pied wagtails are 'fidget' birds, constantly wagging their tails (perhaps not a surprise given their name) and bobbing like a dipper – they are so full of energy. Their constant movement merges their outline into the current of the water and flowing weed; the river's light and dark spots hiding and protecting these little birds with their black and white plumage.

As I walk over the old hatches, I cannot help but marvel at the quality of Victorian engineering still in use today. It is amazing to think that most towns back in the day had their share of heavy goods, from engineering, brewing, and brick companies. The Victorians knew how to build!

The river Kennet rises from springs five miles upstream, around an ancient monument called Silbury Hill, bubbling out of the ground of the Marlborough Downs which act as a huge chalk sponge, before flowing through a paddock at the end of my garden.

The aquifers unfortunately dry up by mid-June, and as I write this there is a wonderful brown trout of approximately 3lbs trapped by falling waters under the bridge at West Overton. Maybe I will catch it and release it further downstream at Marlborough. Flowing west to east, the river enters the River Thames at Reading but many people do not appreciate how beautiful a natural chalk stream it is further upstream above Newbury.

In fact, I would question whether there is anything more beautiful than a chalk stream in May. Water brimming with waving and weaving ranunculus, desperately pointing their starry white flowers with their yellow centres up to the sky, while being dragged around in a merry dance by the clear waters. It seems that Mother Nature has cleverly created the smart, grey wagtail with yellow vermillion to flit and float for flies among the wandering fronds of the ranunculus, perfectly camouflaged in those little yellow centres, as they pirouette their way across the surface.

Deep gravel grooves separate the plants, and the dark of the bankside sedges hides the blackest of 'bumblebee' moorhen chicks, all legs and feet and splashes. Being so shy, the moorhen never strays far from cover, unlike the mallard mother who confidently paddles around while her flotsam of fluff is attacked from all angles: the aerial crow, the stationary heron, and the lurking underwater pike. Luckily, mother duck cannot count, or motherhood would be a miserably sad affair. This abundant fecundity is founded on the multitude of insects and their associated larvae at the bottom of the food chain, all dependent on that everflowing clean, cool, chalk-filtered water. If we are to do one thing, we should protect the water and nature will sort out the rest. It is clear to see that the continued policy of cheap food and water at all costs is now coming home to roost. I would question, how does a foreign investment entity buying into our water companies, and focused on 'ROI' appreciate the nature that is here today right in front of me?

Very rarely do we look at the whole picture: water is the very source of life yet instead of addressing the fact that the valley's arteries are being infected with phosphorus and sewage, we spend our time worrying about the odd habitat here and the odd mink trap there. Whilst important, these aspects are the not the very life thread that keeps this ecosystem alive and healthy, but it's cheap and easy that is where our attention is focused.

On top of this, who is fully aware of the polluting of our riverways and our precious chalk streams? Who really knows, besides a few 'nerdy' environmentalists measuring the 'gammarus' numbers and those of us fishing for trout in our floppy hats who regularly spend time in these waters? The beautiful, bright sparkling water is but an illusion, hiding the relentless decline of wild trout, grayling and stickleback. There is the immediate impact of oxygen depriving sewage, as well as the longer insidious impacts from water abstraction and chemical residues in the water. That said, with natural and <u>extensively</u> grazed water meadows, the Kennet should be relatively free of agricultural run-off. If only the salmon could get this far to find it. The associated reed beds of the Kennet are still a stronghold for cuckoos and their hosts the reed warblers, as well as being a prime stopover location for migrating sandpipers and ospreys.

I have read recently how there is a special wild trout in the Upper Kennet called a greenback – they are quite rare, but I think I had the good fortune to have caught one in 2017. If it were a dog, it would be a bulldog; if it were a rugby player, it would be second row. It was caught by 'dapping' a large Grey Wulf in one of the weir pools, whereupon it shot up from the bottom and did me the honours.

When the Kennet trout throw themselves out of the water. Photo: kind permission by David White.

A week later, the weather has improved, and I am now searching for rises in the river Test. There are no trout showing on the surface, so I peer through the layers of flow, squinting my eyes for shapes hanging, holding sway in the current. In the clear crystalline chalk streams, you may think that this would be a simple task, but trout have figured out that everyone and everything wants to eat them, so their spots keep them well hidden – they are not there for our admiration any more than a tiger's stripes in Uttar Pradesh. Put fish and tigers in their natural habitat and they disappear. Sunlight and polaroid sunglasses aid my endeavours as I scan the shingle bottom for small torpedo shadows giving away the ethereal creatures that seem made of the very water themselves.

As you may well know already, fishermen will often proudly show you boxes and boxes of flies and I am no exception. However, when dry fly fishing on the chalk streams, 95% of the time I only use a Grey Wulf size 8, 10, 12 or 14 depending on the season. There are no large mayflies around, so I start with a size 14 flicked just upstream of the torpedo, to the left, to the right, and then plonked on his very nose with a little splash. The torpedo shows no interest, not even a tiny flicker of a fin, so I move on to find the next shape.

Upstream, being shallower, the current fastens and so does the reaction of any trout. I cast upstream and a fish appears from the shadows to inspect my fly; it follows the fly downstream as it is taken by the current, but just fails to grab it. I repeat the cast, the fish comes up again, studies the fly, then glides back below. I wait a few minutes and then drop the fly vertically above its head. The fish shoots up with intent and grabs the fly. I lift the rod and the fish is darting among the weeds. The Grey Wulf does it again.

Upstream, my friend and fellow fisherman, Jonathan shouts. His rod is bent over creating an arc and there is a lot of splashing in the river. I walk up to see all the commotion, just as he lands a trout complete with large teeth marks across its back. A hungry pike had missed his lunch!

If there is one thing I enjoy more than trout fishing, it is fly fishing for pike. I quickly switch rods and start casting a sinking line with a heavy fly into the river, towards the water under the tree on the other side of the bank. Nothing happens, I feel I am fishing too high up in the water. With the next cast, I push my rod tip into the water, letting the line and fly sink lower into the depths. I repeat the rhythm of long strips followed by a wait, long strip, wait. There is definitely somethings there, but as I run out of time with my line all pulled back through the rings, I finally spot the magnificent terror of the Test glaring motionless at my now placid fly. Frustratingly, once pike have had time to study a big fly they very seldom take; and in this case, the big fish drifts back effortlessly into the dark to lie in wait for its prey.

If you are new to discovering nature, a May morning would be the most enlightening and illuminating place to begin, especially with our native birds. There are the year-round local inhabitants; summer passerines[3] who are still flying through our little Wiltshire paradise on their journeys north, as well as the summer migrants who are here to stay.

On one such morning, I walked up to the West Kennet Long Barrow to the sound

---

[3] *A passerine is any bird of the order Passeriformes (derived from the Latin word passer meaning 'sparrow' and formis '-shaped') which includes more than half of all bird species. Sometimes known as perching birds, passerines generally have an anisodactyl arrangement of their toes (three pointing forward and one back), which facilitates perching.*

of skylarks and spotted my first yellow wagtail of the year. Whilst inside the ancient barrow that was built during the Neolithic period (archaeologists date the long barrow to approximately the 37th Century BC), I spotted two pairs of swallows making their little mud and straw nests high up in the ceiling. An inspired choice. I wonder if they were here last year and whether they remembered their ancient abode when they set off from Africa?

With the exception of the mistle thrush who switches on their 'silent mode' when they have young in the nest, every bird is now in full song. The warblers' sound bursts from the bottom of hedges, while chaffinches call higher in the trees, and from the evergreens the sweet 'squeeze' of the greenfinch, as well as the lovebird bullfinches who mate for life.

West Kennet long barrow on a May morning, looking South towards Devizes

CONSERVATION IN ACTION
GARDENING FOR WILDLIFE

While all advice on gardening for wildlife is useful, the most basic guidance you could follow is to simply to not over tidy the garden. This can miss the point as people enjoy gardening and it is that enjoyment we want to nurture and encourage in a bid to create the diversity of habitats that makes gardens so important for wildlife. So yes, plant bee-friendly flowers, and mow the lawn less, but if you want to get more scientific and provide an even more wildlife-friendly garden, you have to dig a little deeper and be more specific for individual species. Here are two observations for you:

Long-tailed tits love the fat-balls on our bird table but how do we encourage them to nest in a safe environment? These little birds prefer thorny dense bushes for cover. In this vein, I have been tightly pruning a red berberis in our garden to create a better habitat for them, and now we have a long-tailed tit nesting there, as you can see in the image below. I have also ordered three spikey gorse bushes to plant alongside, so I can create the perfect long-tailed tit nesting ecosystem. It is deeply satisfying when your efforts pay off and you see your garden alive with all these wonderful creatures.

Every bird has its niche, and this is true of the mistle thrush, which, each year, has its set nesting routine. First, they locate a large horizontal branch, preferably in one of our mature larch trees and at least four meters off the ground. Timing is everything. By completing the nest in late March, the trees' canopy will be bursting forth just as the chicks are hatching, protecting them from the elements and predators. Mistle thrush like good all-round visibility when incubating, so they can launch attacks on any potential enemies. The parents to-be become extremely aggressive and noisy as they fly to intercept any passing cat, squirrel, magpie or crow. Unlike the song thrush and blackbird fledglings which hop on the ground begging their parents for food, the young mistle thrush chicks sit more patiently high in the trees while they complete the growth of their tail and wing feathers.

We do not all have the space or time to grow mature trees, but if you are lucky enough to have some larch trees, keep them for the mistle thrushes and look out for them each spring. By understanding individual birds and their unique patterns, we can be more specific in how we attract them into our gardens.

This is one of my favourite nests within which I found the long-tailed tit. One of their favourite habitats is this spiny red leaved 'Crown of Thorns' bush.

Is there a friendlier summer sound than the constant chatter of house martins?

…or a more adorable sight than a newly fledged goldcrest.

TAGGING SALMON SMOLTS

For the 'Save the Spring' programme on the river Dee, we began the capture of salmon smolts. Our objective was to catch 100 salmon smolts in the river Muick, a tributary of the river Dee, to grow them on to maturity in a year or two. Upon reaching maturity, these fish can then be returned to the exact spot of their original capture, and hopefully go on to spawn that same year.

Diagram: With kind permission by Atlantic Salmon Trust.

This practice is similar to 'head-starting' curlews, where we look after the chicks in a bid to improve their survival rates, before releasing them as adults. For the curlews, this safeguarding lasts only a few weeks, removing the threat of foxes and crows. For the smolts, this head-start lasts two years, during which time the scientists eliminate the dangers faced by the smolts during their journey to Greenland and back again. These are 100 very pampered and very important young salmon, critical to the survival of their unique genetics, specific to the river Muick.

In May, the rotary screw trap (pictured above) is anchored midstream when the salmon smolts naturally migrate downstream on their journey to the sea. By placing the barge midstream, fish are drawn safely into the revolving helical central funnel, through a tube and into a holding box.

The rotary screw trap in operation.

Note how this salmon smolt has 'silvered up' from its original golden brown parr stage before it's migration to the sea.

Our delight is then opening this box and peering into the mix of salmon, sea, and brown trout, all tails and fins. We release both the sea and brown trout, while the salmon smolts are taken onto the bank in buckets. Here they are measured and weighed – on average they are typically between 110mm and 150mm long and between 10gm and 40gm in weight. A small DNA sample is then taken from the tail and a tiny PIT (passive integrated transponder) tag is inserted just behind their dorsal fin. Combined, the PIT tag and DNA sample will allow us to both track the fish, most specifically to its spawning grounds, and to then identify its successful progeny.

WAITROSE FARMING 2024 CONFERENCE

As a result of both my lobbying for a 'Nature Pathway' qualification to help make the new GCSE for Natural History a success, and my work for the charity Curlew Action, I was invited to the Waitrose Farming Conference 2024. This conference is part of a bold initiative by Waitrose to be net zero by 2035, and their mission is to include all their supply chain partner farmers on this journey.

The conference was hosted at their Leckford Estate in Hampshire, with the theme being the regeneration of soil. The keynote speaker was American farmer Gabe Brown, who recounted the story of how he had inherited his family farm in North Dakota from his father-in-law and how he has brought it back to life upon realising that the soil was completely impoverished. Gabe explained in depth the negative effects of global warming that he is already witnessing on his farm, and how this has been exacerbated by poor quality soil that is depleted of organic matter. This soil no longer absorbs rainwater efficiently, leading to an increase in flash floods, quickly followed by drought. Along with water, soil is one of the cornerstones of life on earth, the foundation of the food chain. Yet so often it is overlooked, sometimes even referred to as 'dirt'. How far from the truth!

According to Gabe, here are the six principles to creating and nurturing healthy soil:

1. INTEGRATE LIVESTOCK: Livestock farming often gets a bad rap. Of course, farmed intensively, beef farming has negative consequences, particularly when huge swathes of rainforests are being cleared for beef and/or soya production to feed the cows. However, if you want healthy soil you need to incorporate livestock, but do not farm them intensively. The major benefit of this is that through the act of animals grazing on plants, it stimulates the plants to pump more carbon into the soil, driving the nutrient cycle by feeding biology.

2. CONTEXT: Be aware that every farm is different and there is no one answer. Regenerative farming is not about a blanket demand that everyone must do X or Y – our natural world is nuanced, and as a result our approach to farming must be nuanced too.

3. MINIMISE DISTURBANCE: Every time we dig, plough, rake or harrow the soil we release more carbon. Synthetic fertilizers, herbicides, pesticides, and fungicides all negatively impact life in the soil. Tilling destroys soil structure.

4. MAXIMISE ARMOUR: Never leave soil bare as it will release more carbon. A coat of armour in the form of plants protects the soil from wind and water erosion, as well as moisture evaporation and the germination of weeds. If you take your lead from nature, you will see that she always works to cover soil.

5. MAXIMISE BIODIVERSITY: The higher the biodiversity levels, the healthier and more resilient the soil becomes. Taking our lead from nature, she never produces monocultures.

6. LIVING ROOTS: Maintain a living root in soil as long as possible throughout the year, whether through agroforestry, biomass, cover crops, perennial forage or perennial grains. These living roots will feed the soil by providing its basic food source: carbon.

These six points all lead to a healthy energy system, improving water and nutrient content and higher biodiversity.

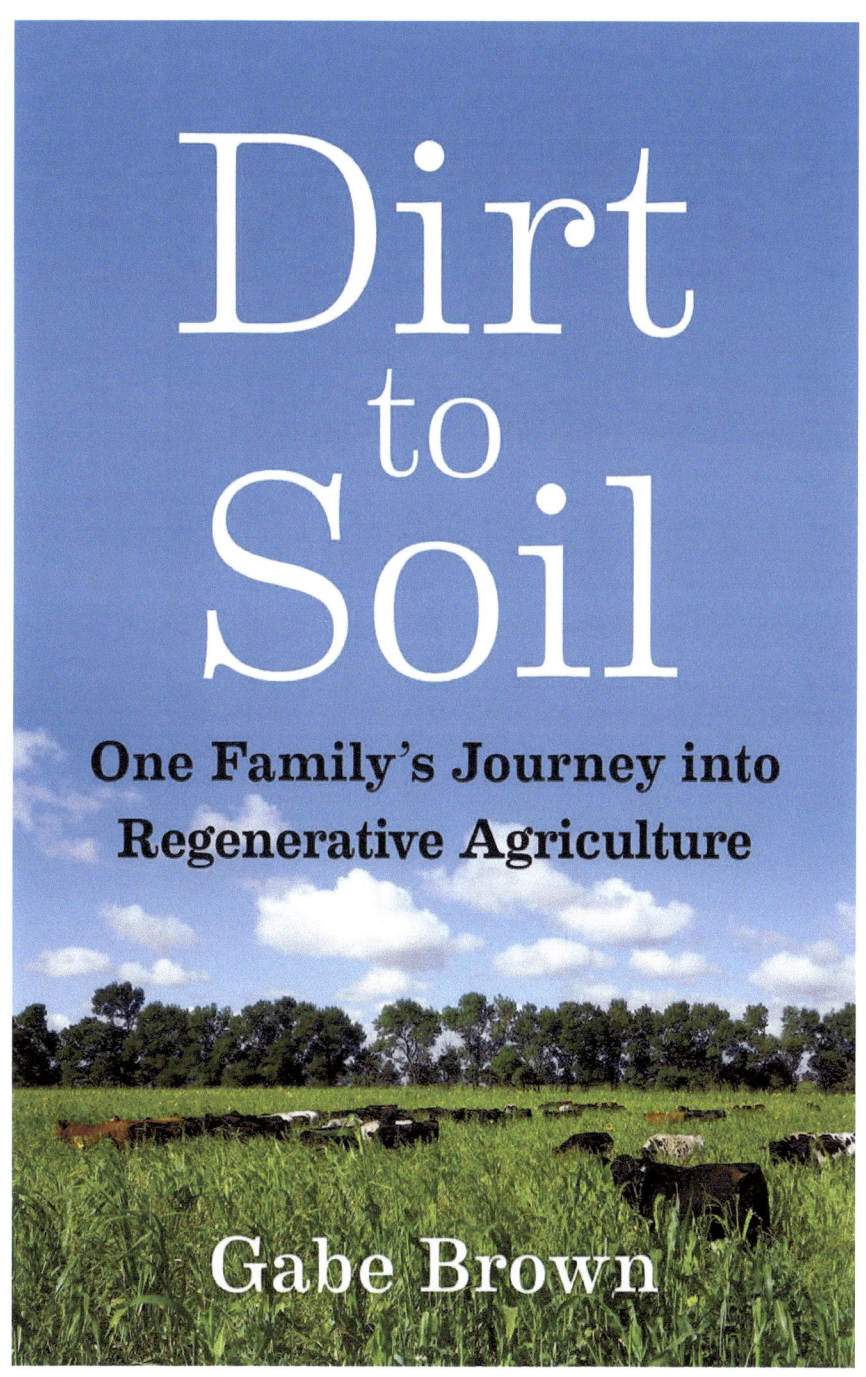

RECIPE: WATERCRESS SOUP (SERVES 4)

In May, the watercress is now peaking in the Hampshire chalk streams. Nature's tonic after the winter. It is a nutritious superfood, full of vitamin C, A, B6 and K, which improves our bone density and resilience against fractures. I love it most for its wonderful flavour. It is best served with fresh bread still warm from baking.

INGREDIENTS

300 gm (three bags) of fresh watercress

Four shallots or two leeks/onions

Three garlic cloves

100gm potatoes

500ml chicken or vegetable stock

Garden herbs two tbsp parsley, thyme, coriander

Two tsp ground nutmeg

Sea salt and freshly ground pepper

INSTRUCTIONS

First, in a large pan sweat a base of onions, shallots, garlic and or leeks, in a mix of 50/50 olive oil and butter.

Meanwhile, in another pan, boil until just tender some new potatoes, or for something different a few chunks of celeriac. Once cooked, add your potatoes and celeriac to the larger pan, together with chicken or vegetable stock, lots of pepper, salt and nutmeg. Find whatever herbs are growing in your garden and throw them in to the mix. I normally use fresh parsley and thyme and then cook for a further 5 minutes until everything is soft and blended. Feel free to add a few peas or broad beans if you have them.

Now the important piece. Add three bags of watercress and a little bit of fresh coriander, stirring all the time for a total of 3 minutes maximum. It is really important that you don't overcook the watercress, otherwise it will lose its peppery freshness. And while we are talking about freshness, try to buy the local in-season cress - there are imports, especially from Portugal, but they don't quite have the same depth of flavour.

Finally, blend the whole mix until you have a rich, smooth and very green soup. If it's a bit thick add some milk – either straight to the pan or when serving it up in bowls.

A heavy shower emphasises the beauty in the fresh growth of water lillies and marsh marigolds at the end of May.

Looking across the River Dee at Carlogie to the Kincardine hut on the opposite bank

JUNE
CARLOGIE AND BALLOGIE

June, a bountiful month that bursts into life with the hedgerows weighed down, laden with Hawthorns' sweet-scented blossom, and ends with the same hedgerows splattered with the beautifully scented, delicate cream plates of elderflowers. Everything is interspersed with the delicate blushed dog rose, and here and there the stronger dominant dogwood.

This is show-time for the British countryside, dressed up in its Sunday best. It is but short-lived and you must indulge it with an open heart, soak up its scents and hear its songs. There is so much to marvel at, especially when you consider how bare and seemingly barren the land looked only four months ago. The millions of birds which migrate here each summer are now back in their breeding places and in full song. I often think, if you met someone abroad and they asked: "What is Britain like in June?" how would you explain it to them? What would catch your eye?

THE RIVER DEE

I have been fishing the river Dee in Aberdeenshire during the last week of June for over 30 years with the same bunch of mates, so every year as the clock ticks towards the end of May the excitement builds. For a fisherman or woman, there is nothing like knowing a stretch of water inside out: high water, low water, and all the flowers, animals, and birds that adorn its tumbling waters. Being June in Scotland, it never really gets dark, so you have the freedom to observe every detail at all hours of the day and night. A quick flashback and I am transported to the constant cries of oystercatchers and sandpipers, the dark green fritillaries on the newly opened thistles, lone otters, and families of stoats tumbling on grassy paths. But ultimately, it is fish and friendship that brings us together, our meeting place.

The Carlogie fishing hut is situated on the romantically named Ballogie Estate on the south bank, grandstanding over a pool called Lucky Hole, (lucky by name, lucky by nature.) The word hut does not do it justice, however, lodge is too fancy – physically, it is somewhere in between, an elysium to fallen friends. There lingers in its aged Victorian wooden beams a patinated smell of burnt logs and fire lighters, Cuban cigars and rollies, Chateau Neuf de Pape and Tennents lager, 90% cacao and Cadbury's milk tray with the gentle growl of extra mature cheddar and well-used waders. A haunt of exaggerated stories, whoppers - always under Chatham House Rules. A homecoming that a dear friend of ours, the late Richard Newton, led every year since the 1990's. He was our captain, our lodestar. We were just cats bringing home our mousy goodies to bask in his glory, with a side dish of alpha male competition – who would catch the first fish, and most importantly who would land the biggest?

The path from the hut upstream treads a sandy trail across gnarled roots of ancient Scots pines, to Richard's favourite sea trout beat, The Mill Pool. You fish flies called Silver Stoat, Blue Crathie, and Black Pennel, before you stomp back in your waders, in the wee small hours, and up the steps. Entering a fog, adjusting your eyes to the stove and candles, wrinkling your nose to the tobacco, we are all ready to put our feet up and for the stream of stories to come.

The Blue Crathie, an excellent low water fly on the Dee. The blurred white building behind is Dess Mill

Sunday night is our tackling up evening, as under Scottish law it is prohibited to cast a fly until the midnight hour on Monday morning. For salmon, a fourteen-foot double-handed 9/10 weight with a traditional spey line and my go-to a Silver Stoat or Crathie depending on conditions, plus a single-handed ten-foot 9 weight with a floating line and a sunray shadow. To complete the trio: a single-handed ten-foot 8 weight for nighttime sea trout. Rod holders attached to the cars are primed and punched ready for the off.

High water, low water, hot sun, cold rain - I have seen them all and fish to the conditions is key, or if too bright or water too low just enjoy the view. 2023 was a prime example of very low water and high temperatures, further challenged by very few fish. So, we adapt and hunt at daybreak or into the gloaming on the fresh fast-flowing, oxygenated stretches: the very top of the Mill Pool, Middle Gannets where it spews down from Upper Gannets, the bottom of the Flats as it pours into the Pot at Potarch, and the fast flow right at the top of the Boat Pool which now reveals some interesting boulders and pockets, where fish can tuck their heads down. It was this streamy run where I lost a fish - the tale told ever since.

Walking down from the Kincardine hut through the meadowsweet and onto the bank where the old pearl mussel shells naturally collect, you cross a side stream 20 feet across, only up to your ankles but with surprising power as it falls away quickly into the Boat Pool - not to mention how round and slippery the rocks are as you tentatively tread your way across. Safely navigated, I walk across the sun-bleached shingle, glaring white in the evening light. The far bank is my destination, with the night sky quickly draining it of any colour, a dark place for fish to hide. The rocks crunch underfoot. Standing on the bank, I do not want to expose my silhouette to the fish, so I stoop a bit and shuffle. I am a good six yards from the water but dare not creep any closer, so I cast as I stand, set back from the noisy flow, with a fast strip at right angles across the water. I want to stir the fish into a take and flashing a collie dog across their nose is one way to wake them up. It's a fast lift cast, lift cast rhythm, no false casts for five yards. Back to the top, I now change the angle letting the fly dangle with the current, slowing it down so the fish have more time to pounce. And pounce one does, tighten the line, set the hook, and I am playing a solid fish in a few feet of fast rushing water. Back on the reel and time to take control, I stand fully upright and the fish now seeing my silhouette and the source of the danger, immediately flips his tail around. He only sat here to get an oxygen fix and is now going straight back down to the deeper safety of the pool below. Reel screeches and the rod complains with the line bumping over the rocks, too fast for me to follow. Ping! Fish lost, but wow that was exciting.

We are very strict these days on catch and release but I am not sure what the rules were back in 2012 for kissing your capture. This beautiful salmon was 24lbs and caught with a size 14 crathie, mid-afternoon on the 4th August 2012 at Middle Gannets Ballogie.

A heavy night after fishing for sea trout.

The Carlogie Hut is a place of dreams and tales and roasted Wiltshire muntjac!

Prepping for the trip is half the fun!

BIRDSONG

When traveling at home or abroad, it is often the birdsong that alerts you to the change of scene, even with short trips within the UK. I have downloaded a bird sound identification app called Merlin, produced by Cornell University, which works worldwide. When you open the app it shows a picture of each bird it has heard and you can then tap to hear that bird sing again. Sitting by the Dee one evening, we recorded 16 different birds in just 20 minutes. Two interesting facts emerged – firstly, a siskin was singing, which are only winter visitors back home here in Wiltshire which I had not picked up. Secondly, the app picks up sounds on the edge of our hearing ability meaning that every time we opened it, we heard treecreepers and bullfinches that our eyes had not spotted. The highlights remain the incessant sandpipers and oystercatchers all day and all night.

FROM GARDEN TO KITCHEN

The first crops are now maturing in the vegetable garden, and the hens are sporting glossy, healthy red combs as they commence their peak egg production. You may have determined from the book so far that I probably find myself more at the carnivorous end of the omnivore scale - and you may have a point - but I would always put quality over quantity with meat, and my perfect meal is a no-fuss grilled steak with a fine salad. My kitchen garden inspires me to cook, and my cooking inspires me to grow. I see vegetables and salads as an extension of my hunter-gatherer psyche. If there are not enough vegetables ready quite yet for a full-blown meal, you can make a seasonal frittata, by just taking the first of the offerings.

JUNE

RECIPE: SEASONAL FRITTATA (SERVES 4)
INGREDIENTS

Six Eggs

One tbsp olive oil

Two shallots or onion or leeks

Two garlic cloves#150g new potatoes

50gm garden leaves (spinach, chard, beetroot, garlic scapes)

Two tbsp chopped garden herbs parsley, thyme, marjoram, coriander

50gm sliced garden greens courgette, runner/French beans and/or mangetout

A cup of peas or broad beans

Sea salt and freshly ground pepper

INSTRUCTIONS

This is a simple, quick, healthy meal, very similar to an omelette but less eggy, utilising whatever you are growing. It has the added benefit of being a one-pan-meal, as you add ingredients depending on how long they take to cook.

First, fry some garlic scapes in olive oil (these are the flower heads picked off your onions, garlic, or shallots – see the below photo), together with a few finely sliced new potatoes (you can use leftover potatoes as well). Cook until slightly browned on one side and flip over before adding crushed garlic. Once the scapes and potatoes are cooked, add the asparagus.

Once cooked, introduce your peas and broad beans (later in the season you can use French or runner beans and courgettes).

Then beat some eggs, enough to almost cover the mixture, and you have this perfect palette of bumblebee yellow fused around an assortment of beautiful greens, accentuating the individual grazing from the garden. Shake the pan so the yellow mix wraps around the vegetables and add on top leaves such as beetroot, spinach, and chard with finely sliced mangetout, whatever you can find!

Finally, add some fresh herbs (which by only cooking for a short time will help to better retain their flavour). I favour finely chopped thyme, coriander, and a little marjoram.

When the egg is cooked on the underside, put the whole pan under the grill for 5 minutes, so it cooks top down. By adding everything incrementally, all is cooked to perfection. Now say thank you to your chickens!

Seasonal frittata

Garlic scapes (flower heads)

TIME TO SIT AND STARE

I have studied spotted flycatchers a lot over the years here in the Kennet Valley, and a warm June evening is the best time for me to quietly sit and watch their comings and goings in the garden. These unobtrusive birds who may flown here all the way from Namibia favour our row of mature lime trees, stretched along the valley floor, which judging by all the swifts, house martins, and swallows somehow trap or attract an abundance of insects. That is my luck, but what is deliberate are the blocks of wildflowers and uncut lawn interspersed with mown paths, creating a patchwork of habitats. I am blessed with a large garden, but I have planted the oxeye daisies and white clover that are now humming with bees, and which are just coming into their own - their rewards are bountiful and beautiful.

The flycatchers, so quiet and unassuming that I suspect many people may pass them unaware, launch their aerial assaults from a favourite branch to chase their prey over the insect-rich grass. If you listen carefully, you can hear their bills snap as they make their catch: feed well little birds, your population has dramatically fallen over the years and this garden is here to boost your numbers before you fly south in the autumn. The birds nest around the house in creepers and boxes I put up, yet every spring I fret in case they do not return. It is with a sigh of relief and joy that I greet their arrival every year. I garden directly for them and all their cousins, and indirectly for these quiet moments when I can sit and stare.

NO MOW MAY

It may have a ring to it, but I mow in May and then leave the lawn in June and July. That way I remove the lush spring growth, giving the flowers a better chance to emerge from the grass, and with them the insects so essential for our birds, damselflies, bats and amphibians - all of whom prey on them.

A male and female bullfinch appear (romantically they are always together) to find some ripening seeds; and families of goldfinches twitter away. A young blackbird is hunting on the strips of short mown paths, together with the dandiest of green woodpeckers hopping after ants. My lawn is not green, it is a tapestry of tweed and brightly coloured dots of white, orange, pink, and yellow, but most of all, it is alive. I just sit and stare. This is my lawn; it could be yours. But first, you will have to throw away moss killer, herbicides, pesticides, and added nutrients.

There is a lazy therapy amid the insect buzz and goings of a living lawn. I trust they find a welcoming habitat in my garden. It is amazing what you can learn simply by sitting quietly and watching your surroundings – whether with a cup of tea or a glass of wine. For instance, I now know it was the great spotted woodpecker raiding my martin nests (I will think less kindly of them next winter when they climb my bird table); and the sparrowhawk may be a wonder, but the swallows have never come back to my barn since it flashed its yellow eyes and launched its stretched talons at the little brood huddled together for the evening.

 The garden has an amazing early warning system for such thieves as the sparrowhawk. First up are the village swallows, followed by the house martins, in a swirling bait ball of cries to taunt the winged hobbies and hawks. They fly fast and high out of reach around each tree that their nemesis sits in, while the songbirds sit motionless, eyes alert. My bantams stop feeding and make a weird clucking sound, too big to fall as prey but still wary and with a collegiate solidarity with the wild birds.

Let part of your lawn grow wild, the bees and insects will thank you (and in turn, we will thank them for their integral role they play increasing biodiversity and our crops).

I wonder if spotted flycatchers were originally named after their young in the nest, as the parents are distinctly streaked. Here they are nesting on one of the purlins protruding out of our house.

Snapdragons, linaria and foxgloves reach a crescendo in late June.

Yellow rattle, the keystone species to a successful wildflower meadow by weakening the more dominant grass.

Hopefully, you will be able to make the most of the early summer evenings that June offers. It is a joy to sit outdoors, breathing it all in, with a cool drink in hand. You may go past the phase of wanting a cold lager or the crispness of a glass of rose, and what could be more refreshing than a gin gimlet? It's been lost somehow in the mass of new gin flavours and contemporary cocktails, but it's the perfect citrusy pick-up around 6pm.

RECIPE: GIN GIMLET
INSTRUCTIONS

Put into a cocktail shaker a 2:1 ratio of gin with a 50/50 mix of freshly squeezed lime juice and some syrup such as honey or agave.

Add ice and shake. To add a sparkle, you can rub some lime around the glass rim before laying over some caster sugar.

Serve with homemade ice cubes embedded with mint or basil, then add twist of lime, and a straw.

I will always love this simple, small and elegant wild dog rose over any cultivated variety.

Showtime for the greenhouse

JULY

WE ARE WHAT WE EAT

With July, comes the first real bounty of the kitchen garden. The fresh garden peas have been in flower for a month now, and at last, the green pods are firm and bold with that sweetest of vegetables you cannot find in any market. Most are consumed there and then, with very few making it to the kitchen. My highlight are the first new potatoes, which used to coincide with my first sea trout supper before scarcity deemed them more valuable alive than on the plate. The herbaceous garden is also coming into its own, with the pink phlox so loved by the yellow brimstone butterflies, foxgloves for the bumble bees, teasels with their towering cones of purple for the peacock and tortoiseshell butterflies and willowherb to attract elephant hawk moths.

FISHING: FOR THE SOUL AND FOR THE STOMACH

Now, just as vegetables are not just about the cooking, so fishing is not all about the fish. It is a rich contradictory mix of solitude and company, of banter and deep philosophical conversation. And so, it was one morning in early July that I took a mate to the river Kennet. Since the fish here are habitually late risers, the plan was to first have a leisurely lunch at the old pumphouse overlooking one of the weir polls: a historical and recently restored Victorian building straddling the river, filled with the roar of water beneath one's feet and birdsong up above. We would then slowly make our way upstream and leave late afternoon.

Loaded with homemade quiche, a steak and kidney pie, and Scotch eggs still warm from their morning baking, we chatted about the various characters and organisations trying to help save the Atlantic salmon. This is serious stuff, the deeper philosophical type of conversation. There are many animal and plant species to worry about, but salmon are on our doorstep and yet, until recently, very little has been said or done about the fish farms in Scotland that cause so much damage to both the fish and their surrounding ecosystems. Increasingly, more people are becoming aware with wider press coverage and much needed campaigning, but we need to do more and with more urgency. We discussed Wildfish and their lobbying of the Government and the water companies concerning the unhealthy state of our rivers; The Atlantic Salmon Trust and their project tracking the migration of smolts (young salmon or sea trout, approximately two years old that are getting ready to migrate out to the North Atlantic from their fresh water breeding grounds); and The Game Conservancy and Wildlife Trust studying the River Frome catchment. It is all fantastic work they are doing, but could The Missing Salmon Alliance be doing more to bring these organisations and other bodies, such as the Angling Trusts, all together to align their efforts in a more efficient and collaborative way? There are a lot of battles to be won with landowners, environmental agencies, the food industry, and the water companies to name but a few, and working together is key.

Stomachs growling, we decided to have lunch just as the intermittent raindrops started becoming a nuisance. We retired inside the lodge to finish our wine and continue our conversation. Whereupon two more weather refugees appeared and asked if they could join our shelter - and soon four people, all of whom passionate, engaged, and knowledgeable, were occupied in a bid to put the world to rights. Lunch drifted on and it was suddenly 4:30pm. We agreed we had indeed had a good day's fishing and carried our rods and exceptionally dry flies back to the car.

Looking downstream from the pumphouse on the river Kennet at Eastridge

RECIPE: CITRUS CELERIAC, FENNEL AND REDCURRANT SALAD (SERVES 4)

INGREDIENTS

One celeriac

Two lemons

One fennel bulb

One cup blanched peas or broad beans

Half cup of fresh apple juice

A handful of redcurrants or pomegranate pips

Sea salt and Freshly ground pepper

INSTRUCTIONS

A great seasonal accompaniment to fatty meats such as pork, duck, lamb, or a fully marbled ribeye steak, is a raw celeriac and citrus salad - it cuts through the heaviness of the fatty meats and adds some zing.

If you buy celeriac the day before, you have the pleasure of its smell emanating from the larder encouraging you to cook. Make sure, if possible, it is fresh, as they do sometimes sit on the supermarket shelf for a while, and you are not cooking this one. Instead, grate it raw into a bowl, before squeezing the juice of two lemons all over the grated celeriac. Add a finely chopped fennel bulb and briskly blanched broad beans or peas. For colour and texture, add a handful of redcurrants or pomegranate seeds. Finally, sweeten with some apple juice. It's simple, quick and gives you all the vegetables you need and may surprise people as being a little unusual.

THE TESTWOOD POOL

It is sea trout month on the famous River Test. A favoured spot is the Testwood Pool, where the river finally spills its way into the English Channel. Twice a day she fills with the salt from the Solent and then, as the tides recede, gallons of fresh Test water pour into the bowl, enticing the migratory fish. A beautiful aquarium, full of salmon, sea trout, bass, bream, carp, and monster chub. Seeing is believing, and it is a little like you are dreaming when you pull your weighted nymph in full sight up past a salmon's nose. He turns, follows, and just as he seems poised to suck in the fly, he flicks his tail leaving a slight boil on the surface.

It is a hot bright day and oh how I long for some cloud to give the fish the confidence to bite with intent, instead they are toying with me. Never mind, there is the evening yet to come with its cooler air and the prospect of sea trout.

Fishing for salmon all day and then sea trout until 3 am allows you to really immerse yourself in your surroundings and soak everything up. There is time, time to watch the grey wagtails feeding their second brood of fledglings, time to hear the whistle of a kingfisher announcing its impending arrival as it flashes upstream through one of the hatches, and time to peer into the deep, through the chub and bream looking for the king of fish – the salmon. Something below is lighter and more ghostly than the coarse fish, but the mouth is a little light in colour and big, and there is a dark spot on its gills. It is a bass, a wonderful bass. I question whether I should switch my fly to a fish pattern and go for it, but no, I am after salmon, and I do not want to spook any which may be destined for my hook. Peering into the deep pools, I am momentarily in another world.

The hot day continues, and I find a pod of about ten grilse (salmon that have returned to fresh water after a single winter at sea), steadying themselves against the current in a back eddy by the top hatch, lazily hanging downstream. There is a little water vole-like plop, as the weighted nymph drops into the water and slowly descends towards these amazing animals fresh from the Atlantic. Once on the bottom, I lift the rod with a steady movement and the salmon jostle out of its way a little spooked and uneasy - time to rest the rods and catch up with Rob.

We drop down to the pool below the bridge, which, with the tide in full retreat, forms a decent flow to cast your fly, allowing it to swing around and then strip across the intersection where the back eddy starts. Fish must be expecting food to be washed down in this maelstrom and a dark shape appears and disappears. I cast again, strip faster - I can see its mouth open, but it misses. I try a slower cast - nothing again. Back to a faster strip and a 3lb sea trout nails the Red Francis fly.

It is best now to rest the water until after dark - as well as our legs in preparation for the night shift - and so I light my portable barbeque. Cooking in a bag of tin foil with freshly clipped thyme, new season crushed caulk white garlic, and olive oil are quartered beetroot pulled this morning from the ground. Added to the mix are some sweet Polish pickled dill cucumbers. However, the main aroma and sizzle comes from two sirloins, nicely caramelising over the charcoal flame. Add wine or whisky to taste and you will be set up for the night to come. Finally, a delicious bowl of freshly picked 'Lord Derby' red gooseberries. This is a large, late variety with a fantastic crispness to the fruit, that seems to pop when you bite into their soft interior. The most underrated fruit but like many things they need to be fresh, so best to grow your own where possible.

With supper over and the light fading, we leave for the beat below the bridge before the tide turns. Falling through the Himalayan balsam and clutching at water mint, I tentatively slide, walk, and climb my way down to the shingle of the riverbed, and fleetingly put the head torch on to check the depth. All good. Carefully, I take each step, aware of the constantly shifting substrate beneath my feet - I want to cast under the trees on the far bank where the sea trout are making glopping noises. Nothing moves and I venture a little further out. The tide has turned though and I need to be quick, but I can just make out a gap mid-river between some oaks with an easy back cast. Cast by cast, little by little I let out a longer line straight down the river. My senses are alert for water, tides, holes, and boulders. A jolt on the rod wakes me from my rhythm as a sea trout grabs my fly, pulls my line, ups my heartbeat and just as quickly spits it out. I cast again quickly, he may have just missed it. Nothing. I cast again, slow retrieve, fast retrieve… nothing. He has gone and is probably upstream of me by now. The water is now lapping my knees, time to retreat.

Back on the safety of dry ground, Rob and I split onto different sides of the main pool that is now filling fast with the incoming tide. It is hard to hear over the sound of water. A large sea trout jumps, and cartwheels, three times right in front of me – I think I can hear Rob shouting, but I take a look and all I can see is the small red glow from his cigarette on the far bank, he's just having a smoke. But I hear him again, and realise it was him shouting and that this athletic fish dancing in front of me is the very same fish he now has on the line. I hurry around with the large landing net. What a leopard, all spots and anger. This is the prize for night fishing at Testwood.

ROEBUCK STALKING

When the first fields have been harvested and you can smell the grain and straw in the wind, it is time to stalk the roebucks again. In the still-to-be-cut tramlines of the wheat, their neat heads poke above the golden sway or from the margins of the field. We spot a head and raise our glasses - it is a doe, maybe with this year's calf hidden close by. We leave her in peace. A movement on the left reveals a pricket (a yearling buck) chasing a different doe.

Chucking some chaff into the sky we check how the scent is blowing and work out a plan. The buck has settled down now and is grazing about 200 yards away, but we will have to approach from the opposite side of where we are standing to be able to take a shot upwind.

A beautiful 7lb leopard spotted Testwood sea trout

We slowly re-tread our steps backwards, heads down, until we reach the hedge we have just come through. Now out of sight, we can make a large steady arc as we cross two fields before entering the buck's field from the opposite side. The first milestone reached so far unobserved. I can see the buck's head lowered to the ground grazing 250 yards away. We are downwind but in full view.

There is a slight contour to our right, a sort of gulley. If we crawl a few yards, we can then walk, crouch, and creep into range. Crawling along the ground, I am cradling the rifle with the full weight of my body in one hand as it pushes into some incredibly prickly field thistle. I just about manage to muffle my groans as I roll away onto my back.

Composure regained, we carry on until we reach the 150-yard strike point. Back on all fours, we edge up the slight crescent, lying flat on our stomachs, hoping the wildflowers and grass will disguise and cover our outline. I bring the rifle around, poking the barrel out in front before I bring my cheek to the stock. A little adjustment of my left leg and slight heaving up of my shoulders and the scopes search their spot. There he is, head up, looking around stock still as I move the crosshairs in for the kill.

All of a sudden, he's gone, off in a bound, then a canter. I look up from the sights and see an older buck chasing him away. Lucky for him but a good stalk, nonetheless.

Being the end of July, there will be many more opportunities to stalk the stubbles.

Willowherb is a great foodplant for elephant hawk-moth caterpillars

The aptly named fox and cubs (Pilosella aurantiaca) where the buds are the cubs and the open orange flower the vixen.

AUGUST

LAXA AND VANHALLE

August gives occasional hints of autumn to come, the odd gale and increasingly shorter days. There is a calm and a stillness, as if Mother Nature has completed her marathon and is now proudly resting, satisfied and content with her work. When the sun shines, it is clear summer is still very much here but I am reminded not to take the month for granted.

One of the highlights of the month are the butterflies – August can be one of the best times to see all the garden butterflies out in force. Buddleia, known as the 'butterfly bush', is a must for gardens, particularly as many butterfly species are now at their optimum populations with their second broods on the wing.

There is, however, a little sadness in August, as the swifts are now departing the UK once again to journey back down south. Of all our summer visitors, they perhaps spend the shortest time on our shores, arriving in late May and leaving by early August. Only the cuckoo may spend less time here, by forcing its chicks onto foster parents. We think of birds like swifts and cuckoos as ours, but in truth, they are away more than they are here on our shores.

SEEKING SALMON IN ICELAND

Conversely, it is also the time I go north to Iceland in search of salmon in their short summer window.

Expectation is heightened once you touch down and hit the N1 highway North of Reykjavik. Driving across the country, the scenery is spectacularly beautiful – wild, rugged and dramatic, it is far from the beauty of the chocolate box Alps, but is no less magnificent or breathtaking. We eventually pick up the Laxa Nordura River after it spills out into the sea at an inlet near Borganges.

Arriving at our destination, are the telltale stationary SUVs, parked randomly at haphazard angles along the shingle bars on the riverbed, all with their empty rod holders. We park up, full of anticipation as we ready our rods, thinking already of the casts we will make into the "strenguir" streams for running salmon.

You learn to read a river and learn where the salmon lie: the confluence of two rivers, the boil caused by a boulder, the tail of deep pools from waterfalls ('fosse' in Icelandic). Each year different fish search the same places where they can put their nose down and let the current effortlessly hold them in place, saving their energy before their journey continues. Your fly needs to cross their field of vision and hopefully make them leave their spot to intercept it. A slight disturbance of the water shows their attempt to bite; a gentle lift of the rod, even from the gentlest of takes, sets the hook and the fight begins. That is the fisherman's thrill, lift the rod, and yes you are in!

That inducement of a bite takes many forms, anger at something in their territory, a distant memory of their childhood in the same river, temperature, water flow, and cloud cover to give them the confidence to come to the surface – but either way, it is often quite gentle for such a powerful animal. If they will not take, then a fly stripped across them as fast as you can, may induce a literal wave of anger which is far from gentle and immediately jolts your rod and sends the line screaming. Either way, it is what we love.

The light in Iceland, being just below the Arctic Circle, is haunting, the landscape wild, and the waters clean and cold, - just how salmon like it, and that is all they ask for. We do our best to miss these facts in the UK. We know the science, but we ignore it. Scotland's heritage punches above its weight, with whisky, shortbread and salmon (amongst other things); but how many people realise their Scottish salmon is farmed like a factory chicken, millions crowded into disease, parasite-ridden cesspits, around a quarter of whom end up dying prematurely as a result of disease and parasites? Not to mention the millions of tonnes of fish required to feed the salmon in their farms. As a result, the remaining wild salmon are too rare to knock on the head. Where are our advertising standards?

Note the tiny holes in the tube of these hitch flies which skate across the surface.

And so to one of the great Icelandic guide's, Teddie Erlingssons', river: the Laxa Hrutafordur. It's a small river only 7km long, a series of shallow runs, and waterfalls with the occasional deeper short pool hidden in the black lava. A young sea eagle's slow wing beats help it lazily cross the valley, only to be immediately mobbed by a pair of golden plovers, their haunting drawn-out "peeeeeew" echoing across the chilled air. As a result, a couple of Arctic Skuas are encouraged to join in the fray. The five birds drift over the hill until the plovers deem it safe enough to return to their chicks hiding somewhere below in the flowery meadow. The sounds of an Icelandic summer.

Being a spate river most of the season, the Laxa Hrutafordur spills into the Atlantic with no visible estuary - dribbling into the tide between some boulders, while the salmon sit it out in the waves. However, after a couple of days of rain in early August, they are able to run to the Kejo pool, where they wait in their droves until the next rain, which may not be until September.

I arrived at the Kejo Pool in early August and counted 50 fish shoaled up. Before I start casting, I watch and observe, tracking movements in the water. A slight splash betrays a small grilse nosing the narrow neck at the top of the pool tasting the fresh water. I watch two more fish follow and then it seems the whole shoal, like sheep, decide to proceed and the bigger two sea winter fish push out the grilse and take their place at the front, a slight boil betraying their presence. There then followed a game of grilse first then the bigger fish push them out, then they get bored, and the grilse return, and the process repeats itself. Their urge to run upstream is constant but the water is too low.

I creep unseen to arrive above the pinch of the pool hidden behind a rock. Crouching here at the bottom I drop a short cast, letting it drift down among the fish out of sight before beginning a slow strip back. Feeling a tug, I lift the rod, and, as I stand up the fish shoots back into the main pool, which erupts with salmon jumping all around. My first fish off the Hrutafordur.

We recently fished the River Halle (pronounced "Halta") in North Iceland for two days, but unfortunately with no rain since June this small river was on its knees. However, as broods of ptarmigan ran ahead on our very bumpy dusty track, we spoke of shooting in early winter with Kari, the lease owner and our guide.

I don't think I could have imagined a more magical setting than spending the day walking up the 20km valley. No dogs, vehicles, beaters, buildings - just the sound of gyrfalcons up on the tops, whimbrels, plover, pipits, and godwits in the tussocky valley bottom, plus the occasional low-flying merlin streaking past you after a pipit. All-natural pasture with the odd sheep and a small herd of ponies.

Kari told us how he is allowed to take 20 brace of ptarmigan per annum, which are a traditional Christmas delicacy. I would swap driven pheasants for this small quota of wonderful grouse any day - less is more, and sustainability is yet still more. I feel we sadly lose the value of a pheasant when each year so many million are put down. Whereas here, every single one of these extraordinary little grouse is special: surviving against the odds, wild and free. Occasionally, man gives a helping hand with Arctic fox and mink control, which also benefits the summer migrants, many of which become our own winter visitors later in the year.

River Halle, Skagastrond.

A DOWNWARD TREND

Compared to our rivers in the UK, there is little agriculture or human run-off here in Iceland; predators are controlled and the redds are protected all winter under a sheet of ice. So far, so good, and that is why people pay a premium to fish here, but even in Iceland there is a worrying downward trend of diminishing catches.

The law is changing on seal management and somewhere, probably far out at sea and for similar reasons as to why our UK salmon and sea trout are suffering, something is not working. Icelandic salmon have less distance to travel to the same winterfeeding grounds off Greenland than our European fish, yet fewer are coming back every year. Is Iceland just a few years behind us? It is a concerning thought.

Manfoss waterfall Laxa Asum, no barrier to the Atlantic salmon.

A sheltered grassy knoll is welcome after 10,000 casts.

RECIPES FOR AUGUST:

Returning from the summer holidays and its, more often than not, rich diet, It is incredibly therapeutic and remedial to cook freshly picked produce from the garden, in particular, runner beans and French beans.

This year's wet summer has been kind to their growth, and I have caught the beans before they become tough and stringy.

This is one of the benefits of growing your own – they never get to that shop-bought tough phase where the beans need peeling . I simply top and tail, then cut in long diagonal strips. These beans are not only delicious, but are also rich in iron, potassium, vitamin A and C, protein, magnesium, folic acid, and all-important fibre which is often overlooked.

RECIPE: RUNNER BEANS & FRENCH BEANS DONE THREE WAYS

RECIPE #1

On the first day, I steam the beans for 8 minutes, then drizzle a little extra virgin oil, add a pinch of salt and pepper to season, and finally cover with some flaked parmesan.

RECIPE #2

The next day, I start off by steaming the beans much like the day before, but replace the parmesan with lightly fried shallots, garlic, red onions, peppers, and courgettes - again all from the garden.

RECIPE #3

On the third day, once again I steam the beans before stirring in a can of tuna and drizzling on top some olive oil.

Three simple meals mostly from the garden's bounty – I feel healthier already!

RECIPE: GARLIC, CHILI, LEMON & THYME SPATCHCOCK CHICKEN ON THE BBQ (SERVES 4)

INGREDIENTS

One medium-sized chicken

One garlic bulb

Two jalapeno peppers

Three shallots or small onions

One quarter celeriac

Two tbsp olive oil

One lemon

One tsp chilli flakes

Two tbsp garden herbs freshly chopped thyme

Sea salt Freshly ground pepper

INSTRUCTIONS

If you want a change from red meat on the barbeque, then try this spatchcock chicken recipe. It shortens the cooking time of a whole bird while increasing the flavour.

You can ask your butcher to spatch the bird, but it is very easy to do yourself. Either with a sharp knife or easier still with some kitchen scissors, place the chicken on its front and cut out the backbone from one side of the parson's nose.

Lay out a sheet of aluminium foil, and cover it with garlic cloves, chilis, chopped celeriac, shallots and onions. Then, lay your chicken on this bed of vegetables with the breast facing upwards.

Generously cover the chicken and vegetables with olive oil, salt, pepper, lemon zest, and lots of thyme, before squeezing out the juice from the lemon, topped with some chili flakes. Some people recommend pushing this under the skin, but I find once you break the seal between skin and meat the flesh is more prone to drying out, so I baste on top.

Seal up the foil and put on the barbeque at medium heat.

Once the chicken is cooked you can turn the flame up a bit and open the foil to get a few crispy pieces stuck to the foil.

Towards the end of the month, the dew starts to build up in the mornings and the odd cobweb begins to appear, but you can always bury your nose in the phlox for a morning pick me up – reminding me that summer is still here and to make the most of it whilst I can.

AUGUST

By doing the "Chelsea chop" (where I cut some of my herbaceous perennials back by a third in May) I can extend their flowering season by a good six weeks.

The Alladale Estate in late August – you can see with the regeneration of mixed woodland

SEPTEMBER
AND THE ROBIN SINGS AGAIN

The exuberance of summer has passed, as we enter September betwixt seasons, when the flora and fauna seem to be at peace with one another. For the most part, the birds have finished breeding, the plants have brought their flowers through to seed and the trees have largely stopped growing for the year - it is like low tide. Everything waiting in the slack. The sunflowers have hit their height, bending east in the morning, twisting with the sun, and hanging down a little more with each passing day; while the Michaelmas daisies point their pollen-filled centres skywards surrounded by their ragged purple petals.

ALLADALE ESTATE: BIODIVERSITY, REGENERATION AND REWILDING

"We call it regeneration please, not re-wilding." Explained Innes MacNeill, as he took us on a tour of the beautiful Alladale Estate, an hour north of Inverness in the Scottish Highlands.

The reason for this, he explained, is that there needs to be some element of management: to simply rewild would not work as without any apex predators, there would be no means of controlling populations of certain animals, such as deer. On reflection, this is largely the case across the UK. We are too small a landmass with too many people to rewild; instead, it is now all about regeneration - where nature has become depleted, we will manage it back to health.

THE ALLADALE ESTATE IN LATE AUGUST – YOU CAN SEE THE REGENERATION OF MIXED WOODLAND

The Alladale Estate was purchased by Paul Lister in 2003 and is now a pioneering eco-tourism initiative. By allowing the natural environment to grow back to its original diversity and beauty, Paul is regenerating and naturalising the land.

Beauty is certainly a word that springs to mind as you drive along the river Carron before splitting right at the junction with the river Alladale. The landscape reminds me of a smaller-scale Yellowstone, although in Scottish terms it is a huge estate at over 23,000 acres. Enough for an experiment but not quite big enough to fulfil the dream of its owner.

The overriding strategy of the estate is to reduce the grazing pressure from the current red deer population (evidenced by the few remaining ancient and dwindling native Scots pine and crag, valley-hugging aspen, rowan, birch, and alder). The transformation is outstanding and the potential for increased diversity of flora and fauna is only just beginning. Consequently, a new financial model is taking form among interested landowners – one that is built on ecotourism, smaller-scale forestry, with far less dependence on grouse shooting and stalking. In any case, an estate must wash its face to succeed.

When considering regeneration, thinking big is key as deer and salmon don't understand estate boundaries. Ultimately, we need more cooperation between landowners and a joined-up approach across society to succeed in making a difference on a significant scale. For now, however, there is finally the evidence to say: Look, it is possible, we can do this, but it is equally important we do not disturb the carbon-catching blanket bog and peatlands. As always it is a case of the right tree in the right place.

This is not about blanket non-native dark, regimented pine forests; instead, this is about a natural tapestry where eventually reduced deer numbers are allowed back in.

CONSERVATION, TREES, AND THE SURVIVAL OF SALMON

Open moorland, free of trees, together, with grouse also provide the safest harbours for our breeding curlews, golden plovers, and lapwings. Like so many environmental matters the solutions are nuanced.

What more can we do around the edges? Especially along the rivers and many tributaries, as an environment denuded of trees leaves our waters more susceptible and vulnerable to a warming climate. No trees equal less shade, which results in higher water temperatures and a reduced root system, which in turn increases the risk of flash floods and erosion. Less vegetation results in fewer nutrients and insect larvae, as well as an increasing susceptibility to predation.

Climate change poses a real problem for the Scottish Highlands, as our formerly relatively cold uplands now constantly break summer record temperatures. One has to ask, are these warmer temperatures one of the many complex reasons why salmon have been in long-term decline and why we have witnessed a sudden drastic deterioration in their numbers and health in recent years? There is a fantastic film called Riverwoods, that does an incredible job shining a light on the perilous state of Scotland's salmon and the need to protect the species, explaining how they are integral to the complex local ecosystem. Imagining a Scotland without salmon is a devastating thought.

Evident, however, in numbers were the swallows and house martins swarming around the buildings where we were staying at the Ulbster Arms in Hawick, right in the north of Scotland. Even at these latitudes towards the end of August, their numbers put Wiltshire to shame. Their numbers in large part are thanks to the multitude of abandoned dwellings and farmsteads that offer shelter to sheep, swallows, and starlings, but which add to a somewhat desolate beauty.

Despite the modern world evolving at pace, there are large swathes of the Highlands which remain untouched by modern infrastructure and tourists. There is a timelessness and a nostalgia about this place. It is an incredibly beautiful, yet unforgiving landscape, and I think the old green fishing hut sums up the isolation of the place.

The wind comes straight off the North Atlantic, so we hide in here with a whisky until the wind dies down and we are ready to cast our lines again.

The Laxa Hrutafordur is shaped by natural larval dykes cutting across the river's path, creating a series of waterfalls and deep pools

A REGENERATION SUCCESS STORY

It seems all the odds are stacked against salmon surviving in spate rivers these days. However, it is not all doom and gloom. This is the success story of a small river in northern Iceland, where one man managed the whole system and brought it back to life.

In 2016, when Teddie, a local fishing guide, negotiated a lease with three farmers for a seven kilometre stretch of river in northern Iceland, the run was down to around ten fish per annum and salmon were nearly lost from the catchment. Unlike most of our UK rivers, water quality and temperature were not the issue. Dedicated to restoring his stretch of river, he posed a simple question: how best can we help the fish and breathe life back into these waters?

His first action was to consider how we, the fishermen and women, might be part of the problem. Immediately Teddie introduced catch and release and banned worming or spinning Next, he started controlling the predators – principally, non-native mink using traps and seals, which were shot in the sea (although this has now been banned in Iceland). The reason this was so important was that most of the fish were concentrated in just ten small pools and as a result the parr and smolts were particularly susceptible to attack. Interestingly, the kelts (spawned salmon returning to sea) stay under the ice all winter in these pools, and I have been told that the mink manage to find access via an airlock and swim down to take chunks out of the torpid creatures.

Seals are a problem since the small river has no estuary and, as such, it simply spills into the open sea, where seals hang around at high tide in August and September. As usual with natural predators, it is the final straw in an already depleted resource that causes the problem.

With human and animal pressures reduced, the next step was to increase the amount of parr (infant salmon) in the river. This may be considered controversial by some but, as I have already mentioned, since we have negatively impacted the fine balance of nature, we must carefully manage what has been damaged back to health. In order to boost the numbers of salmon, aiding the rate of reproduction is vital, so Teddie stripped some fish of their roe (eggs) and milt (sperm), which were then mixed in a bucket, and placed in gravel above a natural waterfall barrier. By doing this, he increased the overall spawning area and holding capacity of the river.

Teddie stripping salmon eggs to help spread them across a wider catchment.
Note the ice along the river.

The result after seven years is a run of approximately 100 salmon and according to the Icelandic Government's fisheries department, based on water capacity, a small surplus can now be harvested which will reduce competition - both in spawning terms and the amount of food available for the fry and parr. As this is a very small river, it is estimated that only fifteen quality hens are required to keep the population sustainable.

HOW THE RIVER SUSTAINS LIFE

When I first came across the river, I couldn't believe it could hold fish of up to 15lbs. Standing at the mouth of the river in the summer you would question if there were a river there at all. However, last week we walked and fished the whole 7km in one day and got to know and understand the river's secrets and its unique ability to sustain life.

The first pool we came across was shallow with little flow and no obvious salmon lies: "Cast across to the grass bank opposite," said Teddie, in a window no more than four feet wide. Sure enough, a salmon of about 8lbs leapt out of the water and took my fly. It was followed by three more and then it all went quiet. Walking across the river we looked at the grass bank and the deep undercut which gave the fish the sanctuary they needed to survive only one km from the sea. No undercut and there would have been no fish.

There were no more grassy banks as we walked upstream and, it became increasingly rocky. By checking each pool it soon became clear that without the pots below the waterfalls, there would be no capacity to hold fish. Rich, dark and hidden oxygenated water: these pools provide somewhere for the salmon to wait until the rains arrive and it is time to spawn.

Most of the waterfalls here are the result of a hard volcanic dyke cutting across the valley - and in one series of three pools, each only a rod length squared, were half a dozen fish. The two deepest pools had large overhangs and you could look down and see the black volcanic gravel cleared of any algae by the constant movement of the salmon jostling for space in and out of a large protective rocky overhang. But perhaps the most extraordinary pool was a mere four-foot by four-foot – one half of which was a bubbling cauldron of dark water and the other half a slower back eddy. The ability of the salmon to survive here, and the speed at which they see and take a fly and leap out of the water was amazing. But then these fish are survivors, we just need to give them a helping hand.

The adult salmon can only survive in the tiny pots of deeper water at the bottom of these waterfalls, waiting for September rains.

A typical Hrutafordur September salmon.

RECIPE: FRESH SALAD DRESSING

INGREDIENTS

One tbsp good olive oil

One tbsp freshly squeezed lemon juice

Two tsp honey

One tsp granulated sugar

Sea salt and freshly ground pepper

INSTRUCTIONS

Mix equal parts of good olive oil, freshly squeezed lemon juice, and a couple of teaspoons of honey stirred or warmed so everything is mixed into an emulsion.

Add salt and pepper, plus a tiny bit of granulated sugar to give it a crunch. Note the sweetness should come from the honey, not the sugar.

This dressing works for all salads, and for those of us who have an aversion to mayonnaise (I have always avoided gloopy sauces like the plague) it is a great way of making a different kind of slaw dressing and goes wonderfully with grated kohl rabi, fennel, and apple, mixed with some crushed fresh coriander. It is the perfect accompaniment to barbequed meat as it really freshens it up with its lightness.

BIRD WATCHING

While the weather remains warm, we eat outside for as long as we can. Every extra day is a bonus in September, and it is a delight to sit and enjoy all the sights and sounds. I regularly pick up the "dui, dui" of a bullfinch - for a bird with such strikingly vivid colours (the deepest pinkish red of any other bird in the UK, contrasted with the pastel grey of a hen harrier and the sooty black head of a crow) they do a remarkable job of concealing themselves. Unlike all the other seed-eating finches, they very rarely come to visit our birdfeeder, which is a shame as I am sure most people are completely unaware of this gem of a bird.

I was convinced a pair nested in the garden this year but, in keeping with their elusive nature, I never found the nest. I see them most often in the winter, especially after snow when they seek out the old blackberry or dock seed heads. Something you will never find in a nature book but a habit I often see now and into the winter is bullfinches feeding on the old nettle seeds. This is another reason to leave a few stragglers uncut until the spring before they move on to feast on the swelling buds of our plum and greengage blossom.

This is also the last chance to watch our house martins until the next year when they make their return. Looking up from a Sunday lunch in the garden and seeing up to a hundred house martins chattering and feeding is a real September special for me. Below them in little trickles and passing more swiftly are the swallows in their gentle curving flight: Bon voyage and safe journey little ones - please come back next year.

A late swift scythes through the air. A heron comes soaring over the high limes at the front, no doubt on his way to our pond, and banks sharply as it sees us, *"kraaking"* as it goes.

CONSERVATION: WHEN DID YOU LAST HEAR A CURLEW?

Experiencing the loss of our diversity of species and the sheer reduction in absolute numbers of once common birds in the UK is extremely sobering. However, there is always hope to be found somewhere. I can usually find some peace, knowing an action group is lobbying and fighting for their existence, a community of like-minded souls out there. We are not alone.

Nowadays, it is hard to recall when you last heard a curlew, or for that matter, a cuckoo, a nightingale, or even a turtle dove. And when did you last see the lapwing's lazy bouncing flight behind a tractor, their sharp acrylic Robin Gilmour patterns now just watery colours etched in our minds across a silent landscape?

We should not be the last generation to have our spirits lifted with the glorious hit, the joie de vie of the bubbling, long curved bill of a curlew forcing winter away with its sheer exuberance. For me, Brechfa Pool near Lyswen in March on the Welsh Borders is my faded, bleached etching. A cock bird shaking the Wye Bluffs with his 'curl lee' statement of ownership and place. One field aloud with the calls of his frothy courtship, back on his breeding grounds for another spring. Or not?

It is mind-blowingly sad to consider that we may be one of the last generations to share our land with these birds but let me offer some hope.

SEPTEMBER

Curlew North Pennines 'back on the moor', photo by Emily-Graham-media.

Ground nesting and in the open they struggle with predation of both eggs and chicks. Photo by Emily-Graham-media.

SEPTEMBER

Curlews usually lay a clutch of 4 eggs, hidden in a tussock.

THE CURLEW RECOVERY PARTNERSHIP

The Curlew Recovery Partnership brings together all those interested in curlew conservation; joining forces to help secure the future of this most iconic and threatened species, the Eurasian Curlew. Their mission is to provide co-ordination and support to those engaged in curlew conservation, while also providing benefits for other threatened species and habitats.

Their overarching framework represents the experience and expertise of nine diverse bodies, to ensure all relevant habitats are represented with a common set of agreed policies.

Why do we have a responsibility?

There are approximately 58,500 breeding pairs of Eurasian curlews in the UK, which is an estimated 20-25% of the global population. However, numbers in the UK have halved over the last 25 years, and there is a real risk of extinction as a breeding species in Ireland, Wales, and lowland southern England.

Over the winter, the number of curlews in the UK swells to an estimated 125,000 birds (62,500 pairs), at around a third of the world's population, but once again, it is a similar story, with conservationists noting a 25% decline in the total number of wintering birds.

"The curlew is now classified as vulnerable to extinction across Europe and is arguably our most pressing bird conservation priority in the UK."

Populations in lowland southern England are now reduced to some 500 breeding pairs, and while with careful management, there are some relative strongholds in certain uplands, we must not become complacent as even here we witness long-term declines.

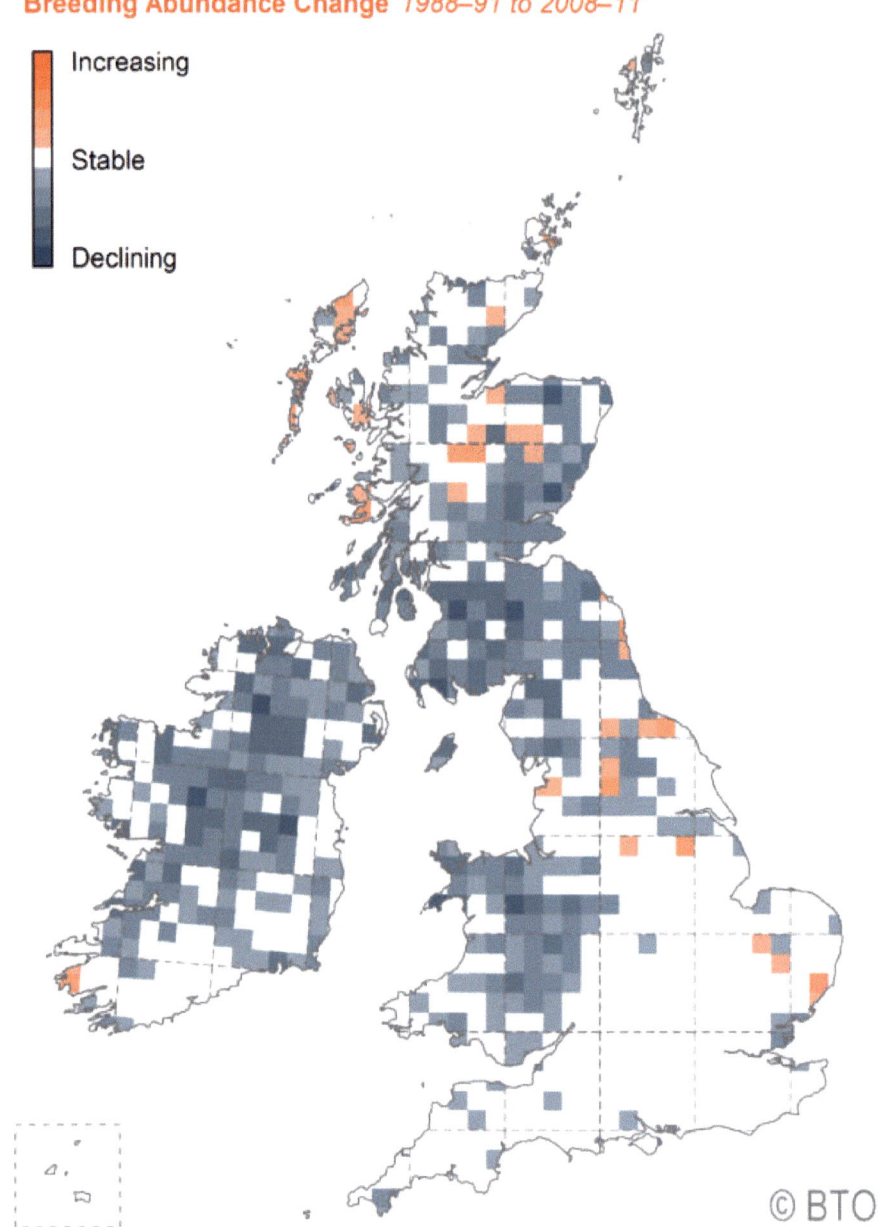

EURASIAN CURLEW ABUNDANCE

Maps from Bird Atlas 2007-11 (Balmer et al. 2013), which was a joint project between BTO, BirdWatch Ireland and the Scottish Ornithologists' Club, and reproduced with permission from BTO.

THE CHALLENGE

We now know that low productivity (i.e. low chick survival rates) is driving the downward trend, as opposed to reduced adult survival – which currently stands at 90%. This has been strongly associated with three factors:

LOSS OF HABITAT:

Habitat loss and degradation include urban development, grassland intensification, drainage, afforestation, and peat extraction.

AGRICULTURAL OPERATIONS:

Some agricultural practices such as early rolling and cutting of grass for silage lead to the direct loss of eggs and chicks.

PREDATION:

Abundant generalist predators such as Foxes and Carrion Crows are key threats at the egg and chick stage.

Providing the right habitat all year is vital, but alone is not enough if chick survival rates are low. To sustain the curlew population, the challenge is to double the current survival rates of chicks from an existing 0.25 per year per pair to 0.5 per pair. In a nutshell, British curlews need to produce about another 10,000 more fledglings per year.

Broadly, you can categorise the necessary actions into three main activities:

1. Managing the current levels of habitat loss, including forestry and recreational disturbance.

2. Mitigating loss of eggs and chicks to commercial grasscropping activities, such as hay and silage.

3. Mitigating the loss of eggs and chicks due to predation – mostly by foxes and carrion crows – by targeted culling, for instance, and nest fencing.

There has been a massive increase in awareness since 2015 when a paper was published in British Birds highlighting the recent declines – this awareness plays a pivotal role in reversing the current downward trends.

Finally, it is worth noting that there is a positive knock-on effect to nurturing the population of curlews back to their former numbers: by aiding chick survival rates and ensuring that the curlew's habitat is protected, you in turn support lapwings, golden plover, redshank, and snipe – all of which were once such a feature of the British Countryside but are now similarly in steep decline.

WHAT ROLE CAN BUSINESSES PLAY IN PROTECTING CURLEWS?

We must all understand our environmental footprint or impact, and increasingly this is accounted for in annual reports as part of a company's overall Environmental Social Governance (ESG) strategy and/or TNFD (Taskforce for Nature related Financial Disclosures). Improving nature is also the key overall apex for the new UK Environmental Improvement Plan 2023 and part of the Environment Act 2021. The new Biodiversity net gain (BNG), mandatory from 12 February 2024, is a new approach to development. The objective is to improve habitats for wildlife in a measurably better state than before the development.

On top of this, shareholders and employees are also demanding more sustainable practices.

Curlew North Pennines 'back on the moor', photo by Emily-Graham-media.

"Future Countryside surveyed in 2023, asking: What makes you proud to be British? Perhaps unsurprisingly, 'The Countryside' came second at 37%, with only the NHS thought of more highly at 51% of the vote."

So, my challenge to businesses nationwide is this: just how authentic are your "green credentials"? Real sustainability requires actions that truly make a difference rather than surface-level gestures. Falling into "greenwashing" traps is easy if offsetting measures don't deliver true environmental benefits. For instance, planting non-native trees in inappropriate locations can harm ecosystems instead of helping them.

Instead, consider supporting impactful conservation efforts that genuinely engage employees, like those championed by Curlew Action and the Atlantic Salmon Trust. While these organizations are dedicated to enhancing curlew and salmon populations, their initiatives go beyond individual species. They work across broader landscapes and watershed areas, promoting biodiversity, ecosystem health, and water resilience.

Through partnerships like these, your business can play a meaningful role in environmental stewardship.

Loch Horrisey photo courtesy Jonathan Evans

OCTOBER

WESTERN ISLES

Every warm and sunny day in October is a bonus, the big and bold dahlias, with their bright and crazy colours, adding drama and an extravagance to the garden - summer's final hoorah. Standing tall and proud, basking in the last days of warmth and defying the cold that is to come. However, the change of seasons is inevitable and when Jack Frost finally appears, the flowers blacken, weaken at the knees and fall over in a matter of hours. The show is over.

These dahlias, snapdragons, marigolds, sedum and verbenas provide a wonderful patchwork of colour as the garden is winding down and getting geared for its winter hibernation.

THE WILDEST PLACE IN BRITAIN

In the Outer Hebrides, during the month of October, the sea trout leave the estuaries and sea pools to run the whisky-coloured burns to reach their birthplaces and spawn.

It can be very wild at this time of year - casting into the wind and pouring rain - but there are also those beautiful still days when you can be alone with your thoughts, with only the sound of geese flying south for company. The calm before the storm.

CASTING INTO THE FLOOD TIDE ('THE POOSH') AT KIRKEBOST NORTH UIST

Standing at the edge of a sea pool in North Uist, I can see the sea trout swimming past, nosing around the seaweed and rocks, looking for shrimp and sand eels. An hour earlier I was on the other side of the pool – if I was there now I'd be trapped by the waters, the tides here change so quickly. You have to keep your wits about you.

To say that North Uist is wild and remote is an understatement. The wind whips up a froth on the lochs and instead of casting we hunker down in the boat as the rain bounces off your wading jacket. Beneath, you have several layers to keep you toasty:

woolly jumper, sweater, shirt, and a vest. Vanity goes out the window as you embrace the 'Michelin man' look – no need to be miserable in the cold.

Like all Scottish fishing, the names are romantic and add to the experience. It is an honour to say I have fished the likes of Loch Horrisay, Skealter, and the moody spiritual Loch Nan Gierann Mill which flows out into the Vallay sea pool, where the haunting abandoned wreck of Vallay House sits brooding on its very own island across the sands.

There is no causeway as such to the island of Vallay, just a wide-open windswept expanse of wader calls, broken shells, and the far away sound of waves. Vallay house was built by Erskine Beveridge in 1902, a lino magnate, originally from Dunfermline. It must have been an Edwardian grand design in its day with water piped across the strand. Beveridge really did live his dream here until he died in 1920.

Tragically, after inheriting the house his son drowned while crossing the strand. The house was then abandoned and has been left to decay over the years – yet, still it stands defiant, suffering the Atlantic gales with no roof, no floors, and no love. Waist-deep in the water, everything you see is analysed, processed, even questioned and the Welsh poet W.D.Davies' poem 'Leisure' often comes to mind:

LEISURE

What is this life if, full of care,
We have no time to stand and stare?-

No time to stand beneath the boughs,
And stare as long as sheep and cows:

No time to see, when woods we pass,
Where squirrels hide their nuts in grass:

No time to see, in broad daylight,
Streams full of stars, like skies at night:

No time to turn at Beauty's glance,
And watch her feet, how they can dance:

No time to wait till her mouth can
Enrich that smile her eyes began?

A poor life this if, full of care,
We have no time to stand and stare.

W. H. DAVIES

Vallay House, now a ruin, has remained untouched since the 1920s following the tragic death of Beveridge's son. Photo courtesy of James Macletchie.
©www.jmacletchiephotograpy and www.igot2travel.com

To me, fishing encapsulates this poem perfectly: **"What is life if, full of care, / We have no time to stand and stare".** After all, one may not catch a fish, but you can find such peace in simply being there in the moment: to enjoy the here and now, to hear the pink foots and see their wavering skeins as they announce their arrival from Iceland. Nature grounds you, the perfect tonic to the pace of modern life.

During one of our outings, we walked up a wee burn where a small dam holds the water back and spotted under the heather a very neat and smooth grassy path into the river: an 'otter slide'. Almost immediately, I saw the headless shape of a large freshly killed eel on the bank, complete with teeth marks around its neck and claw scratches on its flanks. I must have disturbed the otter's meal and it had slipped away, past the dam and into the peaty waters.

Now this was a big eel, who will unfortunately not have the opportunity to migrate back to spawn in the Sargasso Sea. Perhaps it was the dam, plus an old fish trap, that gave the otter an advantage in the hunt. I have witnessed this behaviour before of eating the head first by rats on chickens. The desire to quickly reach the nutritious brains, or is it simply the easiest way to disable and tackle such a large and slippery creature. Carnivores living on single prey items need to balance protein from muscles and lipids from animal fat, so they instinctively seek out first what their bodies tell them they are lacking in their diets.

Similarly, it was only last week that I saw a cygnet in a small paddock near home. I imagined the long grass was preventing it from getting a run-up in order to fly away and thought nothing more about it, assuming it would find a hole somewhere to escape. The next day though, it was lying dead. Perfect in every way, except it had no head and neck and one foot was missing. That is a sure sign of a fox, biting off the easily detachable parts with which it can run away.

When it is only 6C, the fish are nowhere to be seen or hooked, and the sun makes an appearance - what do you do? Find a sheltered nook on a grassy knoll, drink whisky, smoke cigars and talk total bollocks. It is almost as enjoyable as fishing!

NEPHOLOGY – THE STUDY OF CLOUDS

The clouds alone make a trip to the Outer Isles rewarding and enjoyable. I can lose myself staring at the sky as the cold, crisp shapes morph without any visible movement in the stillness of the tideless and waveless loch. The clouds inch their way to sunset and by 17.43 the sun dips below the horizon, bidding us farewell 'til morning.

It is a magical scene: all around me, the clouds are in the water and the water is in the clouds - nature's sea and air wrapped together, creating the perfect cycloramic backdrop. The smell of salt and seaweed is strong. A solitary cormorant breaks the spell with a watery contrail rippling right to left.

OCTOBER

Changing light North Uist sunset looking West 17:24.

Changing light North Uist sunset looking West 17:43.

SALMON FARMING

If it is the rugged outdoors with its stunning scenery and solitude you are after, then the Outer Hebrides is for you. In one day, you can have bursts of glorious sunshine, pouring rain and incredibly strong gales. It is truly wild.

However, like everywhere I go, there is no denying there are fewer fish these days. Over the last 25 years, it is estimated that wild Atlantic salmon have declined by 70%, and as such, the population in Great Britain is now considered as endangered. In stark contrast, there are two things that have increased in number: salmon farms and the sea lice that follow suit. No one wants to hear it, but the inconvenient and uncomfortable truth of the matter is that a lot of us are partly responsible: buying farmed salmon from shops or restaurants is driving the demand - in turn causing the loss of wild salmon and sea trout. The term 'responsibly sourced is a misnomer.

HOW DO SALMON FARMS AFFECT OUR WILD SALMON?

Put simply – salmon farming risks the very survival of our wild salmon and sea trout.

- Forced to swim past Scottish salmon farms on their annual migration to sea, our wild salmon and sea trout smolts risk picking up potentially fatal sea lice infestations and diseases.
- It can take as little as two sea lice to kill a wild salmon smolt (an adolescent salmon), a number that is frequently exceeded around the salmon farms.

Recent studies by the Atlantic Salmon Trust have shown that contrary to previous thinking, tagged smolts do not swim immediately north towards Greenland when entering the sea but instead linger offshore for a few weeks, making them more vulnerable to contact with the salmon pens. The tags are designed to give insights into the complete lifecycle of the salmon, including their return journey to their natal rivers. Hopefully this study will enable us to better understand the dangers that wild salmon are experiencing once they have left our shores.

AN ETHICAL TRADE?

The salmon farming industry is now worth over £2 billion in revenue per annum but at what cost to the environment and the fish themselves? Is there any way that salmon farming could be deemed ethical, irrespective of the harm to wild salmon and the wider environment? Even if we could prevent contact and contamination between wild salmon and farmed salmon, it is a hard case to argue:

- Within the pens, flesh-eating lice – caused by the unnatural conditions – literally eat the salmon alive.

- Mortality rates of stocked salmon have been increasing and the latest figures submitted by the industry itself shows a 4.82% overall rate for October 2023, equating to 13.5 million premature deaths between January and October 2023.

- Cumulative mortality over the full production cycle (this is the percentage of fish that have died on a farm during the entire production cycle) is even higher at nearly 50%. This figure can only be reported once the entire farm has been fully harvested at the end of the production cycle - usually two years.

- From a wider environmental and social perspective, the number of wild fish needed to produce food for farmed salmon (and all farmed fish) is huge and a growing concern. About a fifth of the world's annual wild fish catch is used to make fishmeal and fish oil, of which about 70% goes to fish farms. This is causing problems for fishermen in developing countries, according to the report Dead Loss[4], who are seeing their stocks depleted in order to feed western consumption of farmed fish. Key species such as sardines in West Africa are now heavily overfished for this purpose, and this situation is likely to deteriorate further as the fish farm industry plans to expand in the coming years. Scotland alone plans to double its farming capacity by 2030, while Norway expects a fivefold increase by 2050, according to the report.

A POSSIBLE SOLUTION

Building fish farms in enclosed tanks on land could be a solution to the problem of open net pens, given sea lice cannot live in freshwater. AquaCultured Seafood unveiled proposals for the UK's first land-based farm in Grimsby, a £75 million facility with 50 tanks to rear 5,000 tonnes of salmon a year. However, this proposal is not without its concerns. It is estimated the farm would produce as much effluent as 400,000 humans and that one salmon fillet would need as much freshwater as a human would drink in an entire year. There are no easy answers. The fact is salmon are not the obvious fish to farm with relatively slow growth rates, and a requirement for a rich protein diet.

[4]*Dead Loss: The high cost of poor farming practices and mortalities on salmon farms: www.justeconomics.co.uk/uploads/reports/Aquaculture-Report-v5.pdf*

A Scottish farmed salmon with severe sea lice damage – unfortunately this is not an isolated incident. ©CorinSmith

Nearly half of Scotland's salmon farms burn, dump or destroy millions of dead fish every year, according to Scottish Government data analysed by The Ferret. ©CorinSmith

Standing on the main circular A865 road on the North Uist, even amongst this wild place, when the wind turns, a putrefying, nauseating smell hits your nostrils. You struggle to process this is decaying fish, until a lorry sweeps past carrying the same but a more powerful stink as it transports the dead fish around the island. Imagine the uproar if we did this with cows or sheep.

CONSERVATION & BIODIVERSITY: MAKING A WILDFLOWER MEADOW

Red campion and cow parsley a magical mix

However small your garden, making your own wildflower meadow has moved from fad, to trend, to mainstream. It's a win-win all round for looks, as well as wildlife, but it is not the easiest thing to create.

There are three basic methods, ranked with cheapest and slowest at the top to quickest and most expensive at the bottom:

1. Mow the lawn very short at the end of the summer and dig in wildflower plugs.

Use the lowest setting on your mower, or use a strimmer to scalp the lawn, removing any waste material before planting the plants.

2. Scrape back all the turf and sow a wildflower seed mix.

Lift and remove the turf, allowing you to sow directly on to the bare subsoil.

3. Scrape back all the turf and lay brought-in wildflower turf.

With all three options, you will need to weaken the strength of any grass which can easily overpower the flowers. This can be done by using a plant called yellow rattle. It is an incredibly useful plant: with its combination of low fertility and its natural parasitic properties which saps the strength from grass via its web of roots. This is your magic plant.

Yellow rattle can only be sown as fresh seed i.e. this year's seed before the end of November and requires bare earth so you have to mow any existing grass right down and then rake it out to leave bare patches. This is why autumn is the perfect time to make a wildflower meadow. If you have chickens though, beware of your bantams as they love the yellow rattle seeds!

If taking over your whole lawn is not practical or too daunting then I would recommend starting in chunks – ideally choose a sunny location, and over time you can extend your wildflower garden.

It is important to note that weeding will be required in the first few years for vigorous plants such as dock, nettle, and grasses. However, over time, the wildflowers should take over provided the grass is kept weak. Equally important in maintaining your meadow, is the need to remove all cut material following any pruning in autumn, in order to keep fertility low (you may leave some flowers standing throughout the winter for the benefit of wildlife, but ensure you cut back in the spring). I have now added wild gladioli and spring bulbs to my blocks, and it heaves with insects and voles in the summer.

Mixed seeds are available for all soil types at Meadow Mania *(www.meadowmania. co.uk)* and Wild Flower Turf does what it says on the tin *(www.wildflowerturf.co.uk)*.

OCTOBER

SEASONS OF MISTS AND MELLOW FRUITFULNESS

It is over two hundred years since John Keats wrote his poem 'To Autumn' and that first line still encapsulates the season perfectly as I drift through the apple trees. It is said the young Keats went for a walk behind Winchester College and came home inspired.

TO AUTUMN

Season of mists and mellow fruitfulness,
　Close bosom-friend of the maturing sun;
Conspiring with him how to load and bless
With fruit the vines that round the thatch-eves run;
To bend with apples the moss'd cottage-trees,
　And fill all fruit with ripeness to the core;
To swell the gourd, and plump the hazel shells
　With a sweet kernel; to set budding more,
　And still more, later flowers for the bees,
Until they think warm days will never cease,
For summer has o'er-brimm'd their clammy cells.

Who hath not seen thee oft amid thy store?
Sometimes whoever seeks abroad may find
　Thee sitting careless on a granary floor,
Thy hair soft-lifted by the winnowing wind;
Or on a half-reap'd furrow sound asleep,
Drows'd with the fume of poppies, while thy hook
Spares the next swath and all its twined flowers:
And sometimes like a gleaner thou dost keep
　Steady thy laden head across a brook;
　Or by a cyder-press, with patient look,
Thou watchest the last oozings hours by hours.

Where are the songs of spring? Ay, Where are they?
Think not of them, thou hast thy music too,—
While barred clouds bloom the soft-dying day,
And touch the stubble-plains with rosy hue;
Then in a wailful choir the small gnats mourn
Among the river sallows, borne aloft
Or sinking as the light wind lives or dies;
And full-grown lambs loud bleat from hilly bourn;
Hedge-crickets sing; and now with treble soft
The red-breast whistles from a garden-croft;
And gathering swallows twitter in the skies.

JOHN KEATS (1820)

The village of Avebury in Wiltshire. Photo courtesy of David White Wildlife.

As Keats eulogises a time of harvest and plenty. I look forward to eating nature's bounty; My reward is gently picking an English cox's orange pippin. When the fruit is ripe, the gentlest of touches will suffice - there is a slight softened click, and the apple falls free into my hands, greenish-yellow splashed with red and russet with a slight translucence giving away its ripeness.

I have a ritual for fresh apples: I sit down with a sharp knife and cut a quarter wedge. First the crisp sweet apple is consumed, followed by a chunk of good mature Cheddar, my favourite is the unpasteurised Montgomery from Cadbury in Somerset. A seasonal joy with the simplest of pairings - no sauces, just the crunch and sweetness of the apple and the savoury, nutty softness of the cheese. The perfect duo.

It was a remarkable apple harvest this year, especially considering the drought in June and July, although I suspect the rain came just in time to save the crop. What we did not eat fresh we have had juiced and bottled at 'My Apple Juice' near Hungerford. What a great idea and a great service. Take your apples and collect bottles of fresh juice a week later. Unlike commercial juices, we have mixed in multiple varieties such as cox, and russet plus a few pears.

This is the first year we have harvested our own honey from bee hives in the garden, and I am sure it is no coincidence that our apples have been so abundant due to increased pollination. In addition to more apples, it is thought that due to the antimicrobial properties in raw honey and how it coats your throat, it can help solve seasonal and local allergies. Local bees make local honey from the same pollen we inhale in the air that we breathe, and the idea is that trace exposure to such allergens desensitises us to certain allergies.

On top of this, in an outside shed, we have stored shallots, garlic, and onions which should last us into the new year.

OCTOBER

Garden produce in October in our larder at Gypsy Furlong.

As autumn settles in, the garden yields its final green treasures—chard and Padron peppers, thriving in the lingering warmth. Chard is a great alternative to spinach, especially since it's less prone to flea beetle attacks. Its deep green leaves are packed with nutrients, including iron, calcium, and vitamins A, C, and K, making it a superfood for the colder months. Sautéing it with garlic, a dash of nutmeg, and olive oil until it wilts into a vibrant, nutrient-rich green mush is a simple yet delicious way to enjoy its earthy flavour.

Padron peppers cooking on the brazier.

And of course, we have plenty of herbs that I will keep going for as long as the warm weather lasts. These plants, probably more than anything, inspire me to cook fresh from the garden with their powerful aromas and taste. Still available to cut at this time of year are parsley, thyme, marjoram, rosemary, borage, and mint with lemon grass in the greenhouse.

Left in the ground over winter are leeks and Russian kale. Time of year to take off any butterfly netting from the latter so the birds can find any leftover caterpillars or aphids.

If you want to grow and cook a different vegetable, then I suggest celeriac. Not only does it have a rich earthy flavour, but it also permeates the entire kitchen with its woodland smell when being cooked. A smell that has stood the test of time, comforting you on a wet autumn day. Maybe it is the raw, unsophisticated clarity of unprocessed food that I love, as well as its simplicity.

SINGING OUT THE MONTH

Who is singing in the garden now? Only the robin and the odd late chiffchaff who passes through; but, as the nights draw in, the blackbirds signal the evenings are drawing in with their "chink chink". The days become noticeably shorter and the impending darkening of the evening seems to set the blackbirds off, just as the cock pheasant starts uttering his excited two-tone crow as he flies up to roost in the copper beech tree. There does not appear to be any mutualism or commensalism between them. I like to think they both simply remind each other its bed time, but while the pheasant is bold and brash, strutting his stuff, the blackbirds' song is shrill and nervy.

RECIPE: ROAST CELERIAC

INGREDIENTS

One celeriac

One tbsp olive oil

One tbsp sea salt

Parchment paper

INSTRUCTIONS

Firstly, take your celeriac and cut the top and bottom off so it sits squarely in a pan.

Smother with olive oil and salt, wrap it in parchment paper, and put it in an oven at 180 degrees for a couple of hours, or alongside a roast.

As you will know from this book, I am not one for sauces and condiments. So, once cooked, remove the part-steamed, part-roasted celeriac from the oven and cut into squares. These chunks of celeriac go brilliantly with a rich belly of pork - better than apple sauce which can be too sweet.

If the smell of your roast pork, lamb, beef or chicken hasn't got everyone salivating, then the aroma from the celeriac will certainly be doing the job.

The Viola Tricolor, a field pansy also known as Love-in-Idleness and Heartsease

NOVEMBER

RED LEGS AND FIELDFARES

THE HERE AND NOW

When you first arrive at your peg before the drive, what are you thinking?

Without realising, so often we are reflecting on the past or contemplating the future. But, how often are we truly centred on the 'here and now'? A brief window in the day, a pause, a space between events, where we are verbally - maybe even physically - detached from the people around us. You may be entering a church, waiting to be called to present on stage, or simply waiting for your bus to arrive or for your turn in a queue. A quiet moment of reflection and anticipation.

And here I find myself now, a quiet moment as I walk to my peg before the shoot, a solitary individual focused on the present. A minute ago it was all jokes and laughter, but if you miss the first bird of the drive, the joke is on you. A calm descends - there is a certain rhythm, a procedure, an agenda to follow: First, unglove your gun, then look through the barrels to check there are no obstructions; observe where everyone else is, including the beaters; check where your feet are grounded; have your cartridges to hand; and, then pause. It is all up to you now. All about reacting to that first sound.

No one has shouted "Forward" or "Over" or even called my name yet. The soil is claggy around my boots and fresh raindrops glisten in the dying field pansies. An unexpected purple, yellow, and whiter shade of cream; a summer relic still here, but only just, and if I were not here no one would know, and even now it is my secret. Everyone else is looking up at the grey sky. Instead, I am gazing down at this delicate wild viola called a Heartsease - a flower once used for love charms and often referred to as 'Cupid's Flower' and 'Love-in-Idleness'. The latter seems rather perfect for this moment: I am briefly idle waiting for the call to action, with a deep love of the prescient, and a sharp state of consciousness. I see, I hear, I smell, but don't talk – I love this feeling, yet any second, it will be shattered.

As I scan the hedgerow in front of me for the first whirr of a partridge I am on edge, waiting for the winging across the stubbles. My ears direct my gaze up high in the sky, a "chukakaaaaaaaaaaa" dropping from the clouds. Quickly, a mass of wings and sounds pour onto the red haws in the scruffy hedge. Light grey underwings flashing in the low sun affirm what my ears knew before my eyes confirmed. Fieldfares fresh from Scandinavia or maybe Russia? I wait for the whirr of the gamebirds.

Looking back down at my feet, I see the squelched and compacted clay misshapen by my wellie prints and a crushed, soilstained 'Love-in-Idleness'. The last flower in the field, its moment is gone before the guns start popping.

A few moments later, I delicately and respectfully pick up and cradle the limp body of a partridge - also gone, but saved the dishonour, the ignominy of a misplaced wellie boot.

On the next drive I find myself quoting a little Thomas Moore:

THE LAST ROSE OF SUMMER

Tis the last rose of summer,
Left blooming alone;
All her lovely companions
Are faded and gone;

No flower of her kindred,
No rosebud is nigh,
To reflect back her blushes,
Or give sigh for sigh.

I'll not leave thee, thou lone one!
To pine on the stem;
Since the lovely are sleeping,
Go, sleep thou with them.

Thus kindly I scatter,
Thy leaves o'er the bed,
Where thy mates of the garden
Lie scentless and dead.

So soon may I follow,
When friendships decay,
And from Love's shining circle
The gems drop away.

When true hearts lie withered,
And fond ones are flown,
Oh! who would inhabit
This bleak world alone?

THOMAS MOORE

A TRUE COUNTRYMAN

When did you last meet a real countryman, someone carved into the landscape, part of its ecosystem?

Growing up, we did not have a television at home until I was nine years old. And so, as children, we would run down the road in our pyjamas on a Friday night to Dr. Foskett's house to watch Daktari (Swahili for doctor). It was a weekly series, later made into a film, about a veterinary safari lodge in East Africa with a cross-eyed lion called Clarence.

When finally my father found some money to rent a TV from Radio Rentals, I discovered Jack Hargreaves - a true countryman, not a scientist, just someone who knew what was happening out there in the fields, the woods, the brook, and the hedges.

I mention this because I recently heard Recuerdos de la Alhambra by Francisco Tarrega, which was used as the opening theme tune to the 'Out of Town' series (later renamed 'Old Country') in the 1960s, and so I looked it up on YouTube. There they all are, a treasure, an archive of the countryside in the 1960s, starring of course Jack Hargreaves. Jack was filmed in his home, garden, and neighbourhood in Dorset, simply telling you in his wonderfully relaxed but knowledgeable manner what he was thinking that day or what he saw around him.

Decades before podcasts and reels, he had no scripts, no teleprompt, he just looked you confidently in the eye and spoke. This gave a wonderful spontaneity to his show – he was a brilliant broadcaster and an old-fashioned storyteller who knew his subject well. A perfect example of this is in one episode where he is walking past a newly thatched cottage and notices a pair of starlings making quite a racket on the roof; quietly, between draws on his everpresent pipe, he comments how last year these birds had a nest, but now find it blocked by the new straw, hence the kerfuffle. Simple observations of nature.

Above all, Jack loved fishing. In one episode, he recounts how fishermen knocked the chub and roach on the head to take home to boil in order to provide extra protein for their chicken's 'layers feed mash'. No waste and no food miles. Around 1970, I witnessed something similar as a child at Caversham Bridge on the River Thames. An old boy was catching bleak which he then carefully wrapped in a piece of cloth to take back for his cat. What a lucky cat.

NOVEMBER

The English television presenter and writer, Jack Hargreaves OBE – a true countryman. He dedicated much of his life to highlighting the accelerating distortions in relations between the city and the countryside.

PREPPING YOUR GARLIC FOR NEXT SUMMER

Elephant garlic (the large bulbs) and Caulk Wight drying in the sun after harvesting

I plant my garlic bulbs at the beginning of November – there is a certain satisfaction knowing that part of next season's crops is already in the ground. By planting the garlic at this time of year, I am giving them sufficient time to get their roots established before winter really sets in - providing them with a head start for growth in the spring – and the added bonus is that the kitchen garden will also look productive with their green soldier rows during the winter months. Garlic require some time in the cold of winter, a process called vernalisation, to make the cloves split.

Once grown and harvested, provided the garlic is kept in a dry, dark place, you can continue to consume your crop months after being dug up (I am still eating the garlic that I pulled up in June). When cooking, the beauty of home-grown garlic is that the clove crushes sharply and crisply from the pressure of a flattened knife. Sometimes with shop-bought garlic it is like trying to squash a Haribo or a wine gum.

I recommend growing 'hard neck' varieties, such as Caulk Wight and the so-called Elephant garlic (which scientifically at least is a type of leek technically), as they are more suited to our climate: the stems don't fall over so easily, remaining strong and upright.

HOW TO GROW:

Garlic is very easy to grow provided you have normal free-draining soil, a southerly aspect, and plenty of sun.

Bury each garlic clove about the depth of one and a half bulbs (approx. 10-15cm deep). Any deeper and I find the bulb can grow like a daffodil without the individual split cloves. Space the cloves 20-30cm apart so they have plenty of room to develop.

Once planted, all they need is some weeding and, if possible, some feed and water in the spring, especially if it is dry in late May and early June, as the bulbs are beginning to swell.

HARVESTING:

Harvest your garlic when the leaves start to die back. Gently lift the bulbs with a border fork and leave them on the open ground to dry thoroughly. I find my cooking then becomes increasingly generous with garlic, making the most of its in-season freshness.

RAIN, DROUGHTS AND HOSEPIPE BANS

Two months' worth of rain in November and above average rainfall in October, yet we still have a hosepipe ban in Wiltshire. The reason for this is that the chalk is still soaking up a lot of water after the protracted heatwave of 2022 and the rainfall still needs to percolate further down into the aquifers.

Sometimes the Kennet springs pop in early December, but in a dry winter they may not appear until January. It is the canary in the coal mine, so to speak, as once we get to February, the river needs to be in full flow to survive the heat of the summer.

FRESH FLAVOUR DIRECT FROM THE SEA

November is a great month of the year for seasonal fish and living in the South of England we are lucky to be in range of Pesky Fish – a new fish merchant that opens online at 6pm Monday to Wednesday each week with landings direct from fishermen and seafood producers around the country, although most of the day's catches are predominantly from boats operating around Lyme Bay. Less than 48 hours later this fresh fish is delivered to your door.

The boats that Pesky Fish use to catch their fish are all-day boats, meaning they must come home each night, making them much less demanding on the environment. These boats also help sustain fish stocks by landing mature fish, sourced exclusively from abundant stocks, and caught using low-impact fishing methods. The romantic boat names are typical of the Southwest with titles such as Resolute, Provider, Next Drift, and Clarissa.

In addition to these environmental benefits, you are also supporting UK fishing businesses and communities by paying fishermen the best price for their landings, as well as supporting local employment opportunities in respect of packing and logistic personnel. And then as a consumer, it is wonderful to be cooking and eating the freshest fish possible.

What I find incredible is the sheer number of species we have in the English Channel. Typically, at this time of year each day there will be up to seven species of sole (brill, Dover, lemon, plaice, megrim, sand sole and turbot), plus mackerel, bass, hake, squid, cuttlefish, pollock, mullet, sea bream, John Dory, monkfish, and red gurnard.

I have tried them all, but for me the best for flavour and simple cooking is the sustainable lemon sole. It can stand on its own with a rich flavour not always found in white fish. A delicate fish simply pan-fried for 5 minutes on either side.

The mackerel, I find, are best cooked over the fire pit or barbeque. Being a rich oily fish, they go well with a sweet and sour citrusy slaw made from freshly grated fennel, apple and celeriac mixed with a little honey, lemon juice and extra virgin olive oil - not too overpowering but the perfect companion to contrast the richness of the fish.

NOVEMBER

Fresh lemon sole and squid from Lyme bay Dorset English Channel

RECIPE: BONE BROTH

INGREDIENTS

Enough bones to fill your pan

Beef bones and marrow such as knuckle, oxtail, and shank

OR

Whole chicken carcasses including feet and wings

Six garlic cloves

Two onions or leeks or shallots

Two carrots and or celery sticks

Garden herbs two tbsp freshly chopped rosemary, thyme, parsley

Two tbsp whole peppercorns

Two bay leaves

Two tbsp cider vinegar

Sea salt and freshly ground pepper

INSTRUCTIONS

Forced inside by the shortening days and a crispness in the air, now is the time to rediscover some winter recipes.

The very mention of bone broth hits all the trending buzz words with its natural collagen and amino acids - the 'new' superfood. But we have been here before. As children our mothers knew exactly how to get us back on our feet after a cold, extracting every morsel of goodness with their 'nose-to-tail' cooking. Making a bone broth is an ancient form of cooking, however it was lost for a while in our quest for the choicest, most expensive cuts while leaving the rest of the carcass for the canned dog food industry.

Before you fill your home with holly and ivy, fill your home with the soothing smells of bone broth. Whether the pot is on the stove or in the oven, there is an exudation of goodness throughout the kitchen – the aroma permeates its surroundings, wrapping you in the deliciousness to come.

Chicken broth is supposedly better for your skin and joints, while beef broth is better for your gut, sleep, and mental wellbeing. Chicken takes about six hours while beef over twelve, which may determine your choice - but that's the fun, this is whole-day cooking and whilst the weather blusters away outside, we remain indoors anyway.

WHICH JOINTS TO USE:

- When making a beef bone both, the best bones are knuckles and marrow, but I would also add ox tail here so you can pick off the juicy bits of meat later.
- For chicken (and all poultry) think of the whole carcass, especially the feet which contain the most collagen - a kind of free medicine.

METHOD:

Cover your joints with water and bring to the boil. Simmer for 15 minutes until a scum forms on the surface (I realise this sounds disgusting but that is exactly the point - you are drawing out the impurities from your broth to make it clear not cloudy). Now, pour away the water and place the bones in a roasting tray and cook for a couple of hours until brown and crispy. You are now drawing out the flavour and helping to break down and release the collagen. For the last forty minutes add a few cloves of garlic and chopped pieces of leek, carrot and shallot or onion.

Finally, transfer everything to a simmering pot. Cover the bones and vegetables with just enough water and add what herbs you have left in the garden such as rosemary, thyme, parsley, peppercorns and bay leaves. Pour in some apple cider vinegar (another 'superfood' recently rediscovered), as the acidity breaks down the collagen and makes it more abundant in the broth. You are, at last, ready to sit back and leave your broth for several hours, your good work now complete. Over to the cooker.

For poultry, leave your broth simmering gently for a minimum of six hours; for beef, leave the pot for 12 hours. When ready, strain the liquid in a sieve and place in a cool place, allowing any fat to form a surface film. Once cool, scrape the fat away from the surface to reveal the wonderful jelly underneath – this is your broth.

Once heated, the jelly returns to its liquid form - the more successful your broth, the more jelly-like it will be when cooled.

After straining the liquid, go through the heap of bones in your sieve and pick out any meat - this will be the most tender meat you have ever eaten (hence why I said for beef add some oxtail). When you warm up the broth you can then add some of this meat, plus some noodles as well as chopped spring onions, chili and coriander… your children will think they have gone out to Wagamama!

A BURST OF COLOUR

As November ends, there is one last flower standing its ground: the Guernsey lily (Nerines). Anything going forward is a winter or early spring blossom.

Nerines have quite remarkable colours, ranging from a pale white to darker more vibrant shades of pink and orange, and ultimately an intense magenta. Grow them against a south-facing wall with plenty of light and they will surprise you year after year just when you think everything has gone to sleep.

NOVEMBER

REMAKING THE WILD - IT'S PERSONAL

In November, I gave an update on the river Dee 'Save the Spring' programme at an Atlantic Salmon Trust event at the Curzon cinema in Mayfair—an eventide celebration, presenting the projects and progress of the Trusts' work over the last year. There is profound optimism that we can help the now IUCN-listed endangered Atlantic salmon. With each new piece of data and evidence gathered, we increasingly 'join the dots' to understand how we give them the cold, clean water they require. However, the scale of the task ahead is daunting and should not be underestimated when you consider we are moving at pace from what would now be considered tinkering around the edges to the realisation we need to encompass complete catchments. My piece was about restoration on the river Dee in Aberdeenshire, which I discussed earlier in the book in April and May.

I wanted everyone to understand this is personal and that they should all 'get personal' themselves about what matters most in their lives.

I have fished the Dee continually (apart from one year due to Covid) for over 30 years with the same bunch of mates, and for me, **helping Atlantic salmon is personal.** It is so much more than just fishing.

For years, ' my week on the Dee' has grown in family folklore as my downtime, my brain holiday, and stepping off this world for a few days, all of which make me a better person. What we discuss in the fishing hut is 'Mortimer and Whitehouse' on steroids.

We know every stone and riffle on the water, yet things have changed.

Below my feet, eels would always wriggle away from you, and just upstream, you would step carefully over the pearl mussels – all sadly gone.

We love this place, and we would not be proper fishermen if we were not also naturalists who care about the world around us.

But here is a sobering fact: excuse the pun

In the 1980s, during our week, we would drink around 30 bottles of wine and catch around 30 fish. Last year, we still drank around 30 bottles of wine but caught one fish. So, for the evidence-based data scientists among you, the ratio of one bottle of wine to one salmon has now dropped to 30 to 1.

I reported that the 87 smolts captured in the spring now stood at 82, which is a minimal loss. They have transitioned to a saltwater tank and have grown from the original average length of around 12cm to a healthy 40cm.

In the wild, we might expect 1 or 2 of our 87 fish to return from the sea as adults to spawn, so if we can increase that number substantially, this could massively help natural spawning in the area.

This is a highly innovative approach; we're learning as we go. Saltwater closed containment growth and release have never been executed quite like this in the UK. We're excited to see how these fish fare when released back into the river to spawn.

The same smolts by November. Photo: With kind permission from Atlantic Salmon Trust.

RESULTS GIVE US HOPE

Habitat work is already starting to show potential. I took this picture only two weeks ago, and it shows salmon redds on the river Muick next to one of the 'Large Woody Structures' (LWS) put in by the River Dee Trust. The river has carved out a little bay around the woody structure, creating the perfect spawning environment. As we approached the stream, a salmon effortlessly drifted out of sight, safe under the rootball.

Large woody structure (LWS) on the river Muick. Photo: With kind permission from River Dee Trust.

And here is the same structure back in the summer, showing salmon using the woody structure. Photo: With kind permission from River Dee Trust.

So, It's personal, there is a programme and there is Hope.

SO YOU ARE GOING TO HELP 'SAVE THE SALMON' HOW CONNECTED ARE YOU IN REALITY?

Looking downstream from the bridge where the glide drops into the fast water.

CAPTIVATED – I push my way through the wood's understorey, a few yellow leaves hanging onto life among the scratched gorse, while below, the brackens' dying embers are still bold enough to snag a leg or softly, unknowingly undo my laces. Dawn is quietly in her own time, awakening an intensity of colours, which somehow brings out the smells of the wood as I shield my face from unforgiving whips and branches, which will not yield to my intrusion. Am I the trespasser here? It was unexpected, and I am disturbing the serenity of a 'dead still' November morning. It appears rude, but I have a rendezvous. A secret meeting with strangers from another world. Our paths may have crossed way back in June, when my 'Blue Crathie' swung in the current across their face, causing them to become skittish or simply blend and merge into the deep – none of us will ever know. Destiny is their pathway, I am but a nuisance to the wood, seeking an improbable interception.

Crossing the bridge five minutes earlier, my eyes had followed the perfect calmness of the water to where the stream began to drop out of the pool. The reflections busted by the quickening flow. A rough tumble of troubled waters or Ókyrrt vatn (turmoil) as they say in Iceland. Everything moves downstream, minerals, vegetables, and plants all together until something doesn't and an 'oddity' appears. The 'oddity' was small, a slight bulge in the water, an aberration of physics, healed by the water, only to reappear further to the right close to the bank, followed by another crease, messing with the reflected canvas. Destiny was alive below the surface, and in my way, I wanted to be a part of that. The 'oddity' had brought me down from the bridge, battling in the woods.

The tranquility of the water and the overall serenity of the scene belie the underwater chaos, which punctuates the water occasionally, drawing me closer to my vantage point, now peering through brush and branches.

I am here, eyes transfixed on salmon redds in the river's main stem. I have trapped these fish as smolts for research in the spring before casting flies to their brethren ten thousand times a day in the warmer months across the world. Now they complete their year, journeying back from faraway seas to this very small patch of aerated water, not too fast, not too slow. A foot or two above where the river breaks into a tumble. This is the place – I am their witness.

Peering into another world.

NOVEMBER

The bottom gravels coalesce into a black, green, and brown substrate, stained by months of flow and summer-born algae, crisscrossed with the contrails of caddis fly and stone fly larvae. When the stones are overturned, they shine out a light orange due to the peat in the water, exposing the salmon redds and dotting the area like mini bomb craters. The survival of these beautiful creatures is locked up in these few light-colored patches of scattered gravel. Their childhood in the river, smolt run to the sea and out into the Atlantic and back up the river is all about the 'here and now', right now. **Get this wrong, and a whole generation is lost.** The gravid females feel the danger in these shallow waters. I press against an oak tree to hide my frame and lose my silhouette. I am indeed the trespasser perched on the riverbank looking into another world. I am the intruder.

In one of these orange craters, a long dark shadow lies with her head upstream, her tail and pectoral fins gently holding position, sheltered by the depression she has been bulldozed with her powerful tail. I dare not move as I watch my breath in the cool air drift towards her I instinctively burrow my mouth into the tree's bark.

I long for her to be successful.

What happens next week does not matter. Just give her time now to do her thing and bury her precious cargo. Let her sticky eggs drop into the bottom of the hollow so she can cover them with stones, shoveled with that big fat tail, even if it is the last breath she takes. She has made it this far against so many odds. I am her Guardian Angel glued to an oak tree watching over her.

I blink, and a cock fish now lies beside the hen, materialising from nowhere; maybe it's a ghost. The light is still dim. To my left upstream, the reflections become scratched before bulging into a bow wave heading straight for my hen fish with her head still poised upstream in her depression, her ghost remaining by her side.

I lose sight of everything except a fin, some tails, and an explosion of water like someone throwing a rock into the depths. More bow waves and the stillness shattered. My spell is broken as I stand up proud from the tree. A cock fish had attacked the ghost fish, all aggression, instinct, and testosterone. It was his destiny as well

But as the calm returns and the ripples shift downstream, the dark grey torpedo appears back on the redd, her redd contrasted against the stones. Sooner or later, she will have company, but I must slip away now and leave her in peace.

Cigar Club Winter Quarters where we can plan next year's adventures.

DECEMBER

PULL OF THE TIDES

Outdoors, December can be a quiet month if you are not into shooting, as the garden may be too wet to work and only the mistle thrush still sings his heart out. This is when the pull of the tides often takes me further afield.

SHOOTING ETHICS

Quite rightly, there is much debate about the ethics of shooting reared ducks over the very ponds they have lived on all year. The term, "shooting ducks in a barrel" comes to mind. They have little fear, are confused, and with nowhere else to roost keep flying back over the guns. If we expect to maintain the moral high ground on shooting and conservation, then this thankfully dwindling practice needs to stop. Real duck flighting - the practise of shooting ducks in the early morning or early evening over an open area of fresh water or marsh used by wildfowl when they fly from their daytime resting places on estuaries or large water bodies in order to feed - is the very antonym.

For years we used to flight a small pond just off the River Test. We shot half a dozen if we were lucky, never knowing if Canada geese, mallard, gadwall, or teal would show, which was part of the thrill. It was often at very last light, with ears straining for the whistle of their flight pinions and eyes searching the dark shapes, that you would, if lucky, get a shot in. Sometimes a splash signalled we had missed a duck altogether as it landed right in front of us.

The key to the sport is the moment – a mere split second - when you lock on and swing the gun. No standing and chatting with your mates, instead you are in a constant state of readiness and high alert for half an hour.

The only warning you sometimes get is an early fly-by, high up and out of range, before the birds wheel around and make their descent. That is not the time to look up and show your pale human face or they will be gone. Keep still facing the ground and out of the corner of your eye work out their direction and get ready. A windy night with swaying willow and reeds cancelling out other sounds gives the birds confidence, and they miss out on the high fly-by, coming in sure and straight in range. But make sure you still give plenty of lead, they are always faster than you think with their outstretched necks, and as for geese double the lead.

In the evenings when the geese come in you have the advantage of them announcing their presence from a long way off depending on the wind. That steady honk, honk getting louder and louder as the light dims, and then they are upon you! Good luck!

Occasionally one witnesses the excitement of birds whiffling, when they want to lose altitude quickly, confidently flicking themselves onto their back in order to drop straight down like a stone - you then know you are in for some action.

Waiting in the half-light you see many daytime birds going to bed. Little parties of long-tailed tits chatter constantly, bringing everyone together and the various thrushes make squeak-like sounds as they fly into a sheltered corner, while ravens and crows flap slowly over the woods.

There is a certain irony in our sport that you are afforded the close inspection of some beast you would normally only view in a photo. The drake duck feathers are particularly handsome, be it mallard, teal, or gadwall and this no doubt raises questions from people opposed to shooting how you can kill such a beautiful creature. My reply is that nature, like everything, has a lifecycle; and, treated with respect, individuals looking to shoot and then eat (this is vital) animals that you cannot otherwise buy from a supermarket, will usually have a deep appreciation for the flora and fauna around them and a desire to understand and protect it. Each bird that I shoot has had the best quality life, free to roam, and once dead, is always handled respectfully and eaten; compared to farmed poultry, I know which duck I would rather be, if given the choice!

THE APPEAL OF A MAN SHED

Our house can get a bit crowded, and the snug room with the TV is not really where you want to spend a summer's day or evening watching sport, nor is it a natural home for a cigar. So, a decision was made to build me a 'Man Shed' across the lawn from the house. It is a homage really to 30 years of wonderful and happy memories on the river Dee, a replica of the fishing hut at Carlogie (referred to in June's chapter). Can it be built with 150 years of spilled whisky and puffed-out cigars? Not quite, but I can start work on creating that atmosphere and history this winter! Just like Carlogie, I would like to lounge back on an ancient leather armchair - except they now cost about £5,000 each or exponentially more if the internal sheepwool guts are really hanging out - and smoke a good cigar. I will hang my picture of the river Dee beats and an old tobacco stained stuffed massive sea trout of no name and no river. My perfect hideaway.

Now, building a man shed seems simple enough. Four walls, a roof, a couple of windows, a door and a log burner where I can sit in the winter dreaming about cavorting salmon and hard-to see muntjacs crawling through the blackberries. But of course, it's not just a shed: drawings by an architect are needed to apply for planning permission, and then you need to find yourself a good and honest builder It seems pretty simple from the outside: a wooden structure, timber floor, inside wooden panelling, log burning stove and metal windows. However, the challenge is that it would then feel all modern with no soul, all IKEA without the meatballs and smell of pine like a Jewsons building merchant. I wanted to make the shed look and feel old, circa 1850, with the history, atmosphere and smells to boot: a den of fibs, tales and tails, planning mischief and tricksy trips, where we scheme the next year's fishing trip but with wi-fi, a fridge, and a TV.

The solution was simple really. If you make a wooden shed from old timber, it comes ready imported with old smells. Thankfully George, our builder, not only turned my mood board into a building, but he understood the feel of what I wanted. There were no plans, just a ten minute chat each morning: "Shall we do it this way or that?" "OK sounds good."

Almost immediately as it was finished, my sculptured loving creation became a sports bar, a teenage den, and an outside eatery for everyone – in the process losing its original name and, by mutual consensus, renamed simply: "The Shed". With the whole family taking it over for entertaining, I was then banned from smoking my cigars in it. My mistake, I had been too successful, and everyone wanted to use it.

MATCH A BIRD TO A TREE

Do you associate certain trees with certain birds? I do - if I was asleep on a desert island dreaming of distant shores and fields and hedgerows, my mind would eventually fall on a plump woodpigeon caught by the low late December sun, bathing the pigeon in a slight orange glow, matched by the lichen-covered orange hue of an Ash branch.

In the spring, pigeons love to feed on the opening sprays of ash flowers, but most of all they just seem to enjoy sitting in them - especially on those cold, high pressure, windless winter days with a gently growing mist. While that is an observation, my next association of bird and tree is a tawny owl and larch. This pairing can be attributed to my childhood love of the Observers' book of British birds which always showed owls in fir trees.

This year, I was fortunate enough to replicate my very own tawny owl roosting in a Scots pine tree in the garden. My joy is the photo could be the same picture from the 1960's. December is one of the best times of the year to see barn owls, especially when a night of rain has made hunting difficult for them. Often, all you will see is a blur of white crossing a field or what you think is a plastic bag in a tree, caught in the corner of your eye.

Mrs. Owl got used to me and just sat looking down with her big round eyes – I think I am in love.

Indeed, it was such a plastic bag in the tree the other day that made me stop and reverse the car, whence I spied a barn owl sitting in an old willow next to the river Kennet. There is a deep hole in this tree and I was wondering if it was a potential nest site when suddenly a second owl flew up from the pastures. Clearly, this is a place to watch: perhaps it is simply the owls exploiting a rich food source as the rising river was pushing small animals out of their homes into the fields or maybe they are nesting. Similarly, at the same time, kestrels are quite conspicuous at present all along the valley.

REGENERATIVE FARMING

Having read "Wilding" by Isabelle Tree, I was keen to understand how possibly this could be the future of farming when we have 65 million people to feed on this tiny island. The answer oozed up my wellie boot on 6th December after a night of heavy rain.

The 3,500 acres of Lower Weald clay in West Sussex are a prime example of where we should not farm traditionally and all that was wrong with the old Common Agricultural Plan. Consistent payments for crops irrespective of the geology, climate, landscape, impact on nature or the soil have degraded our countryside across Europe and accelerated a decline in our natural biodiversity. In her book, Isabelle talks about their struggle to make ends meet; I could see why - I was struggling to lift one boot after the next.

As previously mentioned, re-wilding is not the silver bullet on its own and we now talk more about regenerative farming. The answer must be the complete spectrum from extensive to intensive farming and one farm may include the entire compass. Certainly, Knepp pushes towards the least extensive end of the scale, an exemplar of how our farms could be managed according to their location, with the right plan for the right place, not the wrong tree in the wrong place, or even the wrong tree in the right place, and vice versa.

Realising our negative impacts on nature can be quite depressing at times, but here is a farm, or park, of good hope – where, given the space nature will, in her own natural way, fill it. For instance, those lost birds I spoke about earlier in the book are indeed thriving here, such as turtle doves and nightingales, together with important increases in the commoner species and their overall bio-abundance. Roaming 'wild' are Longhorn cattle, Tamworth pigs, and Exmoor ponies with both fallow and red deer, each animal with its own grazing preferences to create a mosaic of habitats and little competition. Too little grazing and eventually the land would turn to trees, too much and it would be akin to some of the New Forest with too little understory. Nature loves the edges.

Knepp has the first wild white storks to breed in the UK since 1415 after some wild birds from the continent were attracted to the Estates' breeding flock, being part of a conservation project in conjunction with the Cotswold Wildlife Park and the Durrell Wildlife Conservation Trust. It will be fascinating to see how these young birds establish different migration patterns. Already, some have been seen wintering in Spain, which only became a regular stork wintering place in the last few decades - influenced by milder winters and new food opportunities, such as the largest landfill site in Europe at Valdemingomez. Previously all European breeding storks flew all the way to sub–Saharan Africa, which is a perilous journey.

The Longhorn beef is mind-blowing

Professor Sir John Lawton, author of the 2010 Making Space for Nature report says:

"Knepp Estate is one of the most exciting wildlife conservation projects in the UK, and indeed in Europe. If we can bring back nature at this scale and pace just 16 miles from Gatwick airport, we can do it anywhere. I've seen it. It's truly wonderful, and it fills me with hope."

Stork nests in the oak trees at Knepp – perhaps this is a throwback to their preference before modern buildings and latter-day cartwheels?

SALTWATER FLY FISHING

Like the storks, the Miller family flew a bit further than the largest landfill site in Europe, and then on some more, past the Sub-Saharan countries to the Seychelles. There is an attraction there for certain fish and my first morning pushed all the right buttons.

The surface calm is punctuated by multiple mouths, ploughing agape the surface layers for algae and copepods, a milkfish larder in the St Francois lagoon. Dimples, sploshes, sprays, and bow waves trouble the sheltered waters as another predator smashes a bait fish. We are making our way to a small knoll that is barely walkable at high tide. And so too are hundreds of bonefish to escape all those lemon sharks and giant trevally that want to eat them. Boat anchored, we slip overside into the warm waters up to our knees and wade to the shallowest point just breaking the surface. 360 degrees around us a dark circle reveals the bonefish and as one's eyes adjust you can pick out individuals competing for the safer shallows. If you are learning to fly fish, this is an insane place. To start, it's warm, the sun is shining, there is no wind and you can see all the fish. On top of this, there are no obstacles to your back cast, while your bare feet are solidly placed in the calming sand. My wife, Dominika casts into the black, waits for the fly to sink, and then starts a steady slow long retrieve. On the second retrieve a small black mass breaks away from the main shoal and materialises as four fish chasing the single shrimp-like fly. A slight increase in retrieve and competition does the rest, as the rod is pointing straight at the fish and the hook set. And Dominika is hooked in the same instance: "This is really good!" Which translates as I thought this was going to be rubbish. A perfect strong silver fish with pointy fins and a rubbery mouth for finding crabs and shrimps in the sand. You can do this all day if you like, it's a sort of fishing heaven.

    That said, we are then after bigger things…

DECEMBER

Finally, Dominika comes fishing although this is more like a fishing heaven – Alphonse in the Seychelles Outer Islands.

The stripey torpedo-shaped bonefish with its spectacular triangular, translucent fins and tail. I think the best way to describe it is a saltwater barbel. Add the fact saltwater fish are generally more powerful and you get a feel for its incredible strength in bending your rod.

GIANT TREVALLY – THE THUGS OF THE REEF

How do you describe the fight with a giant trevally? Because it is a fight – between you and a beast, except the beast is in his element and you are in yours, two worlds colliding.

In his world, he will fight you until you ache, and you wish someone would take the rod away from you. Can you manage to keep holding the rod? It is a moot point almost, you must stick at it - this is what you came for, what you worked all year to pay for, and why you spent 24 hours bouncing around in planes and smiling at border security guards. Am I smiling? No, I am probably grimacing, but in truth, I am just deadly focused on bringing this fish in.

We are fishing what is called the 'push'. Clear cold blue oxygenated water is crashing over the reef, pushing the warm brown inner lagoon water out into a cool trench where the baitfish become disorientated between cold oxygenated water and the hot lagoon, bath-like water. The gites, kings, GTs, and giant trevally understand the tides and lay in wait, or it would be more apt to say, cruise around waiting. We look for rays and lemon sharks coming up from the trench into the lagoon as gites often follow just behind them, ready to pick off anything they disturb. Some rays and a three-metre lemon shark past me, but they are unaccompanied, so I blind cast into the deep just where the warm brown water mingles with the clear blue. I am using a Thomas and Thomas 12-weight rod with a 100lb leader and a 6-inch brush fly with a single hook.

Gary the guide positions the skiff so I am standing barefoot on the prow with the wind coming over my left shoulder so I can swing the heavy fly with a slow big pick-up motion out with a plop. I let the fly sink for 5 seconds, then start slowly with long strips, speeding up as the fly crosses the boundary into the lagoon. I repeat to the left, repeat to the right; check that there are no more rays or sharks coming, and take another shot. Casting blind helps me find my rhythm, as its all too easy to get in a mess from the pressure when the guide says: *"2 o'clock, 30 yards, cast now!"*

I physically tell myself to: *"Slow down John"* before casting. Let the big 12-weight rod and the heavy fly take the strain; slow, steady action and the all-important single haul or double if you can manage it.

Suddenly, there is a flash from the right and a big swirl. I hold the line tight - this is really important with gites. You must set the hook and try to control the fish from the start.

ROUND 1:

Do not 'trout strike' - the rod pointed straight at the fish, left-hand holds the line tight to set the hook. Now carefully letting the loose line catch up, I can raise the rod and we are back on the reel in semi-control. The reel is singing with the pressure on the line, and the handle has just knuckled my hand - two weeks later as I write this, my thumbnail is still bruised black.

I ratchet up the drag so it is super tight in complete contrast to trout and salmon fishing and take control with that 100lb leader to subdue the fish. There is coral everywhere and sharks, not to mention the gites own line snapping jaws. I pump to get the line back, keep the line taught but with a slight curve in the rod. These are barbless hooks: we love fishing, and we also love fish.

Although the drag is ratcheted up and the line has stopped screaming it is all I can do just to hold position. The reel will not wind, I am hurting but I must win.

ROUND 2:

Gary edges the boat forward and I manage two winds, always keeping pressure on the fish. I now have Ernest Hemingways' book 'The Old Man and the Sea in my head' and encouragement from Gary helps: *"Not too much bend"* and *"keep up the pressure, keep winding"*. I am trying to wind, honestly. It would be so easy to give up, what am I doing standing on this boat in the Indian Ocean? But I am here to realise a dream. I have already lost one gite today (broken on coral or the line shredded by his teeth).

I don't say anything but maybe this time is destiny. First, the backing comes into the top ring and I pull my face buff away to help me breathe. Finally, I am pumping the rod and getting a full wind every few seconds and gaining control. I cannot lose this fish, my face may not look it but I am exerting total concentration on everything, calmly talking to myself with a single focus shut out from the rest of the world. A flash of silver and the gite rolls, I can see him.

ROUND 3:

It is 50/50, him or me. Gary is shouting at me to keep him out of the current before the fish makes a last run for it, using the push to fight me broadside on. The fish is then under the boat and heading towards the beach. *"John,"* Gary says, *"it's a 100lb leader, fight him and bring him in!"*

"Back to the prow this side!" I yell. And then, *"No, back other side!"* Gary grabs him and the net is full. The exhilaration is amazing. The all-important tape measure says 87cm, which is a good fish, although they can grow to 120 cm - now that would be insane!

Giant Trevally. Gary said "lift it higher, I said I am trying!"

There are troubled waters ahead with three big black bobbing rays and several yellow gite tails thrashing behind, hunting the terrorised bonefish. We have found the rat pack, a constantly moving seaside terror headed by peaceful rays screening bare teethed gites underneath and behind.

Gary turns the boat so I can take a shot into the swirling bevy of madness. I strip fast, strip straight, strip low. Three dark shapes follow: water boiling, I feel the fish, set the hook, and tighten line. With the fish on, I guide the slack line through the rod. Brake turned up but still, the line whistles through the rings, and then the Shilton reel spins angrily.

I keep saying to myself that I am in control, but all too quickly I am out of control. I then have the realisation that a smaller more agile but beautiful bluefin trevally had nipped in at the last moment and seized the fly in front of the gites' nose.

Bluefin trevally caught on a surf walk, being pounded by waves casting into the breakers.

MILKFISH – THE GENTLE GIANTS OF THE REEF

One of the strangest fish on the flats is the milkfish. Like the reef manta rays, they are entirely plankton feeders, working their way along the surface. Catching them is somewhat technical as you need to find a feeding pod with their heads out of the water. They grow to over a metre long and fight similarly to salmon with their clean torpedo shapes and propensity to leap like Salar Salar. I was lucky enough to fish some this year.

Motoring on a skiff to the St Francois lagoon, we passed through several channels where, depending on the tides, the water was either flowing in *"on the flood"* or out *"on the drop"*. When the moon is half full, we have the neap tides, where the delta between the flood and the drop is at a minimum, which can result in a buildup of algae in the lagoon. This then concentrates as it is squeezed out of these channels, especially just before low tide. It is at these spots that we look for feeding fish. More often than not, it is the mantas that first catch your eye. They swim slowly upstream mouths open, letting the current swamp through their gills to bring them their bounty before perfecting an Olympic-style roll somersault to go back to the beginning and repeat the process. Their broad backs break the water and the *"wing tips"* appear black and shark fin-like, although they are gentle giants in reality.

Where the rays have majesty in their movements with their large mouths constantly hanging open, the milkfish, while chasing the same food, frantically open and shut their mouths, bunched together in pods and long chasing lines. And that is what allows you to catch them on the fly.

While you can hunt them on the open seas outside the reef, it is here through the channels inside the lagoon that you have the greatest chance of an interception, since, like the mantas, once they have found a good feeding spot they will go around and around all bunched up together, providing a more consistent target for the fly. Now your rod may be a 10 weight but your leader is only 20lb with a smallish size 6 fly made to look like a piece of algae. This is careful, gentle fishing. You wait for the manta to pass as the trail of *"milks"* approach from the bottom of the channel, swimming frenetically. The more frothing on the surface and a general feeding agitation with their mouths competing for the same surface area, the better. They have massive eyes to find the concentrations of algae and plankton they seek, covered by a thick transparent membrane for protection.

Standing at the bow of the skiff with the wind coming nicely over your right shoulder you execute long casts beyond and in front of them. Once the fish are almost level with the current, I swing my fly in an arc towards them, purposefully making a long, slow strip aiming to pass the fly right across the open mouth of the feeding fish.

Casting over the milkfish may make them spook, so I find setting this trap the best form of capture. It may take several casts to connect or you may not connect at all but ideally, you will be casting at right angles to their movement.

Eventually on one such swim past, the line tightens and I manage to set the hook before lifting the rod and letting the slack line catch up with the reel. I am now connected to one of the best fighters in the sea and all around are snags of coral. The big fish at first stays with her mates, which gives us time to open the engines on the boat. Once the throttle is pushing the boat towards her, I loosen the clutch on the reel and the fish then bores out of the channel towards the open sea.

This is what you want. To quickly distance yourself from the reef and be in open water. I tighten the drag just a little, but am careful when doing so, remembering that I am only using a 20lb leader and a barbless hook.

This is the point at which the salmon comparison comes in, as the fish leaps a good twenty feet into the air, somersaulting as I keep my rod flat on the surface. It is at this moment that I realise how big this fish is…. and it is big at over a meter long! This is going to take some time. As the line now pours from the reel, all we can do is follow in the skiff.

When eventually the fish slows down, we use the motor to gain a little slack and pump some line back on the reel, but as soon as the connection puts pressure on the fish she is off again, the fluorocarbon singing through the rod rings. This goes on for 30 minutes, which is not unusual for milkfish: you gain a bit of line, lose a little and each time when you think the fight is nearly over, she goes off on another run.

It is now hurting under the hot sun and Yousaf, a true veteran guide of Alphonse offers a drink but all I can do is utter a faint grunt. All my senses are focused on feeling the fish. When can I apply more pressure and gain some ground and when do I need to let her run? I am not losing this fish.

The clutch is as tightly notched as I dare, so each run is now fairly short, but I still need to release any winding pressure when she decides to push. With each throttle of the engine, I do, however, gain a couple of winds and after 20 minutes from her last jump I can now see the broad back, eye and mouth of my adversary.

Except that's the wrong way around, I am her adversary, she is not mine. I am simply in awe of a beast who can tow us around the lagoon for over half an hour. Most fish will tire within a few minutes and once on the surface succumb readily for a quick release. Milkfish do not suffer such lactic acid build ups in their muscles as most fish and she now decides to keep her head down lower in the water.

A bit of a standoff begins. The leader is about 12 feet long and I don't want to bring it through the rod rings for fear it will snag and the fish be lost. I cannot physically lift the rod 12 feet, so she keeps her distance swimming around and sometimes frighteningly disappears under the boat just out of range of the net.

My option is to try and slow her down by keeping her head at the top of the water, where it is warmer with less oxygen, and then as we do with salmon, while standing at the prow of the skiff, gain some momentum, and guide her to the back of the boat where Yousaf is waiting with the net.

Finally, she is safely in the net. I feel the relief and the sheer satisfaction - I can talk now, and just by slackening the line, the barbless hook drops out of her mouth. Keeping her in the water the tape measure stretches to 112 cm, and I am just able to stand on a coral 'bommie' for a quick photo, and peek over her blue-green back before returning her free to roam the seas, with those big alien eyes.

DECEMBER

It was an honour to hold this beautiful milkfish and then watch her massive tail with a sharklike wiggle propel her out of sight into the deep.

Big eyes, big mouth but no teeth! The milkfish is a gentle giant.

FEET ON TERRA FIRMA BUT EYES ON THE SKY

Fishing the flats is a long day, with breakfast at 6:30am to meet by the boat at 7am all sun-creamed up and ready to go; and then back at the pontoon for 5pm. For us guests, it is demanding mentally as you are psyched up all the time. But for the guides, it is sheer hard physical exercise, polling the skiff backward and forwards across the reef, engaging with wind, currents, and tides, all while your eyes battle the reflections looking for fish.

At the pontoon, I thank the guides, clean the rods, and jump on my bike back to the bungalow. I ride past Grumpy the giant tortoise who was originally found on the reef with a rope tied through a hole in his shell. Rescued from this fate, he now lives free with his friends on Alphonse as part of their breeding program.

A quick shower and I walk onto the beach, turn left, stub my toe on some coral, and order a glass of Sea Brew. I now lie back on the sun lounger with my cold beer, cigar and thoughts of the day. Complete peace.

The wind drops to zero and you can hear the mullet troubling the water; and far out across the reef, in the lowering sun, you can see the first specks of the atoll's birds coming home to roost. By far the most numerous are the frigate birds, both greater and lesser species. The former, massive birds with angled wings, hanging like our red kites in the UK but bigger, and all black except for their magnificent red pouch. Then the smaller lesser species with some white on the throat and underbelly.

Before they land in the native tall beach casuarina trees among the imported coconuts, they wheel and soar together at the western end of the island, stacked up like planes waiting over Heathrow or salmon waiting to leap the Falls of Feugh on the river Dee. Among them flying swiftly in singles and faster, come the noddies like smaller, brown versions of our gannets.

The beer and cigar are going down well. I feel I have earned my right today to just lie watching an ornithological bedtime, with birds becoming silhouettes as the light fades. A grey heron drops by, the same as our English birds. Does he have a better, easier life here I wonder? He is all feathers and wings as he bashfully re-arranges himself - until a fish puts him into elegant moonwalking mode.

Most fascinating, I find, are the whimbrels and turnstones which have migrated here from the Northern Hemisphere - the same birds I see when fishing in Iceland. Bizarrely, I come across these birds inland, above the coconut plantation; yet, you would never see any self-respecting whimbrel or turnstone in a beechwood at home. I wonder whether these could be our birds? I would love to know, but, in all probability, they have most likely come from northern Russia.

Something different then catches my eye, not a sea bird but the darting purposeful flight of a falcon. Now, what is this bird doing here in the middle of the Indian Ocean? It flies a big arc out across the reef before stooping low over the surf, banking quickly, then springing upwards, vertically like a teal in the dying light. Watch out little Russian turnstones.

On talking later to the Island's Conservation Officer, Elle, her opinion is that it is an Eleonora's falcon on migration, most probably from the Mediterranean heading to Madagascar. Wow!

Eurasian hobby wintering on Alphonse Island.

However, the next day, Gail, who is charged with monitoring the bird life on the outer Islands manages to capture a photo of the falcon and firmly IDs it as a Eurasian hobby, sending me this information:

"We observed the Eurasian hobby from the beach on the 22nd of December being chased by fairy terns. We assumed it is immature due to the absence of 'chestnut red trousers'. The Seychelles Birds Records Committee confirmed it, and according to the national database, this species has been seen once before on Alphonse in 2016 (another immature) and a total of 34 times within Seychelles. At the moment we have an Amur falcon around, which is much more common. Anyway, until next time – happy birdwatching in the UK"!

The conservation practices on these outer islands show what can be achieved with responsible eco-tourism. Everything is protected, all fish are returned and it probably has the largest concentrations of bonefish in the world together with green and hawksbill turtles. This is sadly a marked comparison with the reduced bio-abundance on the mainland.

The prettiest and daintiest birds here are the fairy, or white, terns. Smaller even than our little terns, with the local name of sea swallow. Again, they are doing very well with their new-found protection, and with a period of calm weather settling in, they are now coming into one of their breeding seasons. I say one, because they lay a solitary egg on the open branches of Cordia (geiger) and Callophullum (mast wood) trees, where the eggs and chicks are extremely vulnerable to high winds and rain. With such a high rate of attrition, they may attempt several nestings before they have any success.

The terns, along with the wedge-tailed shearwaters, have suffered from the island's resident, but non-native, rat and feral cat population. However, the government is planning an eradication process in two years, so these islands can once again be safe harbours among the waves.

MAHONIA: THE STAR OF DECEMBER

Back at home and even now in the depths of winter, a mild spell brings out a few winter flowers in the garden. The bulbs, daffodils, and snowdrops are beginning to show tiny buds breaking through the soil. However, the real stars of December are the yellow drapes of mahonia. Bees will fly on mild days when the temperature reaches a critical 13c or above and my three mahonia will hum noisily with their wing beats as the grass below splatters yellow with their petals.

If you are trying to create a garden with flowers for every month then this mahonia is a must. Early next year blackcaps will eat the berries and the dense prickly coniferous foliage is a good nesting site for many small birds. The blooms from the plant also make a great Christmas table bouquet with one of their flowering spikes set as a centerpiece.

Even in December there is joy to be found in beautiful blooms, such as Mahonia.

One compost heap and 87 grass snake eggs were all successfully hatched; my contribution to the local herpetological community.

JANUARY

CULTIVATING NATURE'S BIODIVERSITY

NURTURE AND NOURISH YOUR GARDEN

Behind every good gardener is a good composter. I like the natural closed loop of knowing exactly what I am putting into my soil (note the words 'soil' and 'garden' which are often replaced in the USA with 'dirt' and 'yard' which rather unfortunately miss the beauty to be found on your own doorstep). Composting is a vital part of gardening: if we are what we eat, then plants are their soil. Hence the importance of nourishing our soil with top of the range, homemade compost. No trips to the garden centre, buying dubiously named 'friendly' or 'responsibly sourced' compost, sheathed in plastic bags, devoid of worms and bugs. Just multiple wheelbarrow trips to my compost piles – who needs a gym!

Together with worms, my 'homemade' compost has improved my soil structure by breaking down the clay and acting as a mulch - holding in moisture and suppressing weeds. Garden compost is not a fertiliser and the nutritional content will depend on how you make it. The fundamentals are that anything goes, but the key is mixing green (e.g. grass clippings) and brown (e.g. fallen leaves) so you don't get a build-up of one substance, especially grass clippings which can become very anaerobic, dark and slimy if congested. Re-cycling your shredded paper works, as well as all non-cooked and non-meat kitchen waste.

On the nutrition part: a word of warning. I mixed barrow loads of horse manure with my compost which resulted in a build-up of phosphorus which was too much of a quick fix for my veg, especially the leeks, raspberries, peas and kale. Plants generally prefer a ratio of 2:1:1 nitrogen, phosphorus and potassium (N-P-K).

The reason for this is that the nitrogen is quickly consumed or leached away and without the 2:1:1 ratio can lead to a surplus of phosphorus which inhibits the plant's absorption of calcium. In turn, this causes yellow leaves and weak stems as the plants become chlorotic. Better, therefore, to stick to plant-based compost if you are adding a layer each year, and if using manure make sure it is well rotted!

GRASS SNAKES AND COMPOST

A more unusual consequence of having a compost heap or two (for those with a garden large enough, I would recommend having two compost heaps as you really want to be able to be add to one, whilst you take from the other) is that you can create the perfect habitat for nesting grass snakes.

Each year, when I remove the compost, barrow by barrow, to the Kitchen Garden I find their old eggs. Like a turtle's egg, a perfect slit at the top shows the egg successfully hatched and proves the compost temperature is working as it should. Last year I found 87 eggs which was probably the result of at least 3 separate females laying in the same place. Now that made me curious.

The egg cases are always at the bottom of the heap, glued to a thin layer of last year's dried leaves. I think this juncture between leaves and decomposing grass above is the key. That way they keep aerated from below and the side with the heat evenly spread from above. They also have a southern aspect, sheltered from the north and east by some scruffy undisturbed nettles and long grass. It must be an exact science for incubation to work and some years they are not there. Maybe those years it's just too cold outside or I have piled up too much grass, with the extra weight squeezing and restricting air flow which lowers the temperature. Finally, I keep disturbance to a minimum. Like most protective mothers, the female snake keeps close to her nest during the incubation period.

OBSERVING BIRDLIFE

January is the time of year that the RSPB runs its annual *"Big Garden Bird Watch"* survey, providing vital data on changes in bird numbers and their habits. The trends identified by this survey are key to understanding the impact of the changing climate and landscape on our birdlife and how best we can help to nurture and protect these species.

On a similar scale, but not limited to January, I contribute bird sightings to the British Trust of Ornithology (BTO) on their BirdTrack app. It's easy to do, enjoyable and interesting as it gives you insights into the wildlife in your garden or any other location you wish to report. You can analyse the data anytime and each January, they send me a summary of the previous year, as the image below shows. The joy for me is knowing that I recorded 57 different species and while some of these will be flyovers, for the vast majority it will be a conscious habitat decision to stop at our garden.

JANUARY

Observing and recording are integral parts of being a naturalist.

Bird song in January is predominated by the mistle thrush, loud and brash with his voice, a lofty chorus, from the tallest bare leaved poplar he can find, come rain or shine. On slightly warmer days, robins will welcome you to the garden and a wren may have a sudden burst of delight. While starlings, if you still have them, hedge sparrows, song thrush, and skylark may also occasionally break through.

THE VALUE OF YOUR KITCHEN GARDEN

As for us humans, we are often a little subdued in January. Christmas has been and gone, along with your turkey and pigs-in-blankets; it's cold, it's dark, and you are probably the heaviest, most lethargic you have been for the past 12 months.

Every year, we beat ourselves up in an attempt to shake some order and productivity in to our lives: taking part in the latest fad diet; joining a gym; attempting dry January; giving up meat; the list goes on. I was once minded to start a new yoga class, but the concrete floor and breeze block walls were like the Arctic. It would have been kinder to snuggle up at home with the latest Netflix series. In fact, why go out at all in January?

And to top it all off, the weather is usually not much better in February, but at least the snowdrops will be out. At this point, spring seems a long way over the horizon. However, I am not one to wish my life away and there is some planning to be done whilst the plants rest, their lazy buds.

So I do my best to lean into it a bit, follow the storm, coach, and distract myself. After all, January is an exciting time to plan the Kitchen Garden for the upcoming season and place my seed orders. It's also a month for thoughtful preparation— where to sow the seeds, identifying what needs to be prepared, planning crop rotations, and organizing the sequence of sowings to ensure a productive year ahead.

So, spend some time thinking about what are you going to sow? What do you want to get out of your vegetable patch?

I can split my reasons for growing each of my vegetables and herbs into four categories:

1. Quality and freshness
2. Nutritional content
3. Flavour and tastiness
4. Quantity of produce and economics

Everything you grow should fit into at least the first two camps. So, with an average of sixteen plant varieties that I sow each year, I begin at quality and freshness, with hopefully a better nutritional content for all sixteen.

I would then call out nine plants where the flavour and tastiness are a reason to grow them in itself: peas; French, broad and runner beans; tomatoes; carrots; beetroot; cucumbers; and new potatoes.

JANUARY

Most of these peas fail to make it as far as the kitchen.

But, when it comes to the economics and quantity of produce, I am down to only four varieties that I can honestly say are value for money: French, broad and runner beans plus new potatoes.

Everything else is more of a luxury, considering the time and effort taken, and as the saying goes: "One for the mouse, one for the crow, one to rot, one to grow". In my garden, it would be: "Three for the vole, three for the snail, three to rot, and one for your pail". That said, I love the whole process and planning my Kitchen Garden gets me through the January blues. And ultimately, the voles attract the owls and the snails the thrushes, so it isn't so much of a loss afterall.

My go-to companies for seeds feature heirloom varieties for flavour, such as Real Seeds who are based in Wales.

A winter visiting fieldfare enjoying the apples I put out on the lawn.

JANUARY

FISHING FOR PIKE

Warmer January and February days may see me on Bowood Lake fly fishing for pike. I think I am the only one. The other anglers are generally huddled around their swim feeders, well-wrapped up in their clothes and thoughts. I prefer to keep moving, casting here and there, hunting along the margins, seeking out the old weed beds on the lake floor.

Pike do not fight particularly well, so the excitement is in the take - the 3-second explosive opening of a gob full of teeth, and the flick of their powerful tail. And with pike, you could be reeling in a 2lbs or a 20lbs fish. The thrill of the unknown.

A pike coming to an Alphonse "GT" fly in Bowood Lake Wiltshire.

Today it is quite loud; guns are lined along the far side of the lake with pheasants being driven from the woods above.

A lone picker-upper and his Labrador are on my side watching for any shots which might carry a bird across the water. And there's one, a nice bird, at 20 feet above an oak tree it folds and the momentum carries it with a crash to the nearside bullrushes. Off runs the dog, all obedience and purpose in pursuit. The reeds rustle and sway, and out he pops, bird in mouth, seeking praise and adulation. It is a ritual.

Back to the job in hand – fishing for pike. I am using a 9-weight 10-foot single-handed rod with a floating weight, forward line and an Indian Ocean saltwater large black and purple size two Giant Trevally fly, will this do the job?

The water has cleared up after the recent rains and visibility is quite good, both for the fish seeing my fly and for me to be able to see what is going on. The air temperature is still only 7 degrees Celsius and I suspect the pike are still feeling sluggish near the bottom of the lake, so I let the fly sink with each cast before starting a slow retrieve, stripping a little and letting it sink again.

After 15 minutes, nothing has happened although I may have seen a shape following my fly, or perhaps that was wishful thinking. I decide to try a fast retrieve after letting the fly sink. Almost immediately a fish follows, but as I run out of puff with the retrieve the pike turns away. This is the golden rule of fly fishing for predators, they will often take at the last minute, so you need to keep retrieving to the end.

With the next cast, I stand further back from the water. As the fly comes in, I lift it in the water, as if it is going to escape and I see a flash below – the pike just missed it. I try the same technique again and bring the fly up with a deliberate boil on the surface at the end. Quick as a flash a mouth engulfs it, and I am on. Yes, the gite flies work even in a Wiltshire Lake.

I carry on covering water and have two more 4-5lb fish in quick succession, but I am really looking for their big mother. The big she-pike then appears, as if from nowhere right by my fly, a ghost, but fails on the hit. Now I know they are there, I am wiser for next time.

For anyone who hasn't fished for, or seen a pike before, I would describe it as an ancient fish, surviving on its ability to have a lightning strike and a massive jaw full of backward-facing teeth that can open in a rush of ingoing water.

This is what fascinates me, the speed of the take. Pike have this large underbite giving them a smiling assassin toothy grin and typical big round predator eyes, quite pronounced on the top of their heads with a ring-like golden iris around the big black pupil.

They are fully aware of their surroundings. They lunge at their prey, powered by a dorsal and anal fin set far back next to the tail so the 3 appendages act in unison like a massive paddle. They are plain ugly, but I am bewitched by the majestic blobs of orange on the flanks which blend into backward slanting bars above the belly and multiple spangled scales over the back. Despite all this, when you look down in the water it is their shadow that gives them away, not their colours.

JANUARY FUNGI

On a late January day, you may be surprised by something visually very loud and scarlet on the woodland floor. It will brighten your day, and although not rare, the scarlet elf cup is always a treasure. I find it in a small damp wood growing on decaying hazel wood, often partly covered by moss.

You have to see it to believe it, but there it is brightening up the day, food for small mice and voles during the darkest days, and then for slugs when the wind becomes milder as it blows in from the west. These elf cups are steeped in folklore, which is not surprising for a such a startling fungus. The elves supposedly drank the morning dew from its scarlet bowls hence the name.

Steeped in folklore, these 'elf cup' fungi can be found in late January if lucky.

JANUARY

THE FUTURE OF FARMING

A song thrush is singing - the first of the spring. It is so good to hear they are still around as I have not seen one all winter. They were such a common sight when I was a child and while their decline in gardens by 50% between 1970 and 1995 is pretty tragic, on farmland they are down by as much as 70%.

There doesn't seem to be a single scientific cause and no doubt there are multiple reasons, but if we continually lavish 'molluscicides' which not only kill all slugs and snails but also take out earthworms, we degrade the soil to such an extent that creatures, such as song thrushes, are put at risk.

Our green and pleasant land is not so pleasant and so many of our creatures cannot find food in crops of barley and corn - they are monoculture deserts. All those green fields that supported thrushes, starlings and lapwings in their thousands are quiet and every creature squeezed into smaller and smaller refuges along rivers, paddocks and uplands where the dreaded agriculture treads more lightly. As a result, our gardens become ever more critical bastions of nature and the onus is on us not to bring standard agricultural practices home.

Habitat degradation increases the threat posed by predators, such as magpies and badgers, and disturbance caused by dogs and cats. One of the most important ways we can help the birds, mammals and amphibians is to try to leave somewhere undisturbed in your garden.

There is so much more we can all do, but a vital act was the banning of metaldehyde-based slug pellets from March 2022. But I wonder how many packets are still lurking in garden sheds, garages and greenhouses?

Leaving the EU presents opportunities and risks but if the new Environmental Land Management Scheme (ELMS) is fully implemented then there is cause for optimism - but that is a big 'if'. We need to act and act with urgency.

The ELMS has three key tenets in England comprised of three payment schemes:

1. The Sustainable Farming Incentive which focuses on soil health and reducing the use of 'inputs' such as fertilisers and insecticides. This scheme is being expanded to include payments for looking after hedgerows, grasslands, and soils.

2. The Landscape Recovery Scheme which will pay landowners for ambitious large-scale 'rewilding' projects including natural flood management, peatland restoration and enhancing woodland.

3. The Countryside Stewardship Plus Scheme which will reward farmers for action to support climate change adaptation and for helping nature.

Provided the schemes are implemented, we will be rewarding farmers and landowners for creating a better countryside overall, with improved biodiversity – rather than just enhancing small pieces of ever-disconnected nature reserves.

Substantial environmental improvements will only make a difference if we involve the farming community. The Game and Wildlife Conservation Trust (GWCT) has been at the forefront of delivering more significant benefits to soil, water, and wildlife at a landscape scale through its 'Farm Cluster' initiative, recognising the limited effect an individual farm could make in isolation.

The pivotal role of their CEO, Teresa Dent OBE, was the driving force behind this pioneering momentum. It began with establishing the first group of farms on the Marlborough Downs in 2012, focused on farmland birds, particularly tree sparrows. This success was followed by additional pilot projects in collaboration with Natural England in 2014, culminating in the first-ever Farmers Cluster Conference in 2017, further solidifying the concept of a broader landscape approach to conservation.

That first song thrush singing in late January on a mild day is a wonderful thing, a privilege to hear. As the weather warmed today, the thrush was accompanied by a mistle thrush, a robin and a chaffinch. A promise of what is to come. The days will get longer…

JANUARY

Look at these photographs of the Slender-Billed Curlew, because you will never see this bird again, ever.

Going, going gone! Photo credit Richard R Porter

HOW MUCH DO YOU KNOW ABOUT EXTINCTION?

Everyone learns about the dodo, a large, flightless pigeon native to Mauritius that became extinct in the late 1600s. Its very name has become a powerful metaphor, serving as a reminder of the fragility of life and the consequences of human impact on nature, synonymous with extinction itself. The phrase "dead as a dodo" symbolises something completely and irreversibly gone, a poignant reflection of finality and loss caused by human intervention.

But who remembers the extinction of species like the Great Auk, with the last recorded sighting on Eldey Island, Iceland in 1844?

The last remaining passenger pigeon 'Martha' in Cincinnati Zoo Ohio, which died in 1914.

And the magnificent Thylacine, the last remaining marsupial wolf, which died from neglect in Hobart Zoo, Tasmania in 1935.

 All these creatures were driven to extinction by man's activities, but for some reason, it's the dodo that most people remember, from collective memory over 350 years ago, before the Internet, digitised records, photographs, radio, television, DNA sequencing, and mass media. All this from a faraway place that only a handful of sailors had ever witnessed, but somehow, this strange creature took on a symbolic life beyond its actual existence. It represented the advance of travel and science in the 17th century, a piece of culture, a discovery, which increased the tragedy in our hearts as well as our heads

Now we reach 2024, and IBIS (International Journal of Avian Science.) is about to declare a European bird, the Slender-Billed Curlew extinct.

Smaller than our Eurasian curlew, about the size of a whimbrel, it was last known to breed in South West Siberia, although previously it had been more widespread across the Steppes of Kazakhstan. Not unlike our own curlews it then migrated to the coastline for winter, centred around Southern Europe and North Africa in the Mediterranean.

The last confirmed sighting of Numenius tenuirostris (Slender-billed Curlew), often referred to as "the slim beak of the new moon" due to its elegant, crescent-shaped beak, was in 2004. Before that, a famous photograph from 1995 shows the bird next to a Eurasian curlew in Yemen, captured by ornithologist Richard Porter. On a Curlew Action podcast, Porter vividly recounts the moment when he stealthily approached a slight ridge near a sewage lagoon in Yemen and describes the bird as noticeably more elegant and delicate compared to the Eurasian curlew, with a smaller, more graceful down-curved beak.

Photo credit Richard R Porter.

And here is the sad, really stupid bit because it is gone now. The primary cause of the decline is thought to be excessive hunting on the Mediterranean wintering grounds. We always talk in conservation circles about how there comes a time when a species' scarcity reaches a tipping point, where predation is the final straw, which ultimately tips it over the edge. We have enough challenges managing the habitat for threatened birds around the Globe, but the Slender-Billed Curlew is a reminder to us all that we, as human hunters, should not be that final predator.

WHY DO I PUSH THIS POINT?

Our own Eurasian curlew is Listed as Near Threatened on the global IUCN Red List of Threatened Species, and of course, it is not a legal quarry in the UK. But it is declining sharply, especially across lowland England and Wales primarily due to a low survival rate for the eggs and chicks, due to changes in grassland management and increased predation.

Some conservationists might be surprised to learn that Curlews have been regularly hunted in France, where they are still a legal quarry species. In contrast, they were removed as a legal quarry species in the UK back in 1981 and outlawed in Ireland only as recently as 2012. Largely due to various regional curlew projects and charities such as **Curlew Action** and the work of the **Curlew Recovery Programme**, which includes the **GWCT**, there is a greater awareness of how we can support this culturally significant bird.

In July 2019, after a public consultation, the French government proposed a bag limit of 6,000 birds, including non-coastal birds. At the time, James Robinson, Director of Conservation at WWT, said:

> *"We believe this decision is an entirely political and defies accepted best practice and the biological reality of Europe's declining curlew population. Whilst we do not deny the rights of French hunters to hunt, we passionately believe this must be done sustainably and in accordance with international laws and agreements designed to protect biodiversity."*

The Ligue pour la Protection des Oiseaux (LPO), the largest environmental non-governmental organisation in France and part of Birdlife International, has also campaigned for several years for the banning of hunting of the Eurasian Curlew, European Turtle Dove, and Black-tailed Godwit. All three species have unfavourable conservation statuses, with the Turtle Dove population decreasing by nearly 80% between 1980 and 2015.

The result, since 2019, has been a moratorium, banning the shooting of all three species of birds in France. This means it must be revisited on an annual basis, with legal protection only temporary.

As we near the end of the shooting season It has got me thinking about woodcock

Before we wring our hands and suck in the air about our French colleagues, we need to make sure we adhere to our laws, and where they are perhaps behind the curve, I suggest we lead more by example, follow some restraint, and at the very least follow good practice. I shot my last woodcock in 2010 and remember cooking it the traditional way with gizzard intact, served on toast to my late mother. "Mum, "I said. "This is the last woodcock I will ever shoot, so enjoy it!"

Today, no need to raise the gun, I am simply honoured to witness a woodcock flushed from a covert. A special sight, the bird's characteristic erratic, "jinking" flight full of unpredictability, head-down posture, as it sweeps between the trees. Being primarily nocturnal, we are offered but a glimpse at best, into their hidden world. A bird that may have recently crossed the North Sea, been hatched in Scandinavia, or even flown from further East in Russia.

Back in Scotland each June, their evening 'roding' accompanies me on two sea trout pools on the Aberdeenshire Dee, on the Carlogie beats 'Pitlslug' and the 'Boat Pool' . As you cast your fly across the water dark in the gloaming, but with the sky still bright above, you are alerted to a crisp and sharp repetitive "chi-sik, chi-sik" as a male flies quite high in a dead straight line across the river. It's always a highlight whether I catch a sea trout or not.

The woodcock was amber-listed in the UK in 2022. After a joint survey by the GWCT and BTO in 2003 and 2013, the UK population was estimated at around 55,000 males, representing a decline of 29% over the preceding ten years.

It is important to note this is a census of breeding birds since our population is substantially increased with winter migrants from continental Europe.

Photo credit Ian Harrison.

WHO IS FOLLOWING WHO?

Ten years later, in 2023, a repeat survey estimated a UK population of just under 52,000 males, an overall further decline of 8% since 2013. However, in this time, a divergence in numbers appeared with further declines, most notable in Scotland, especially Northern Scotland, while encouragingly, England and Wales saw a welcome increase. As I have already said, I will never shoot another woodcock in my life; that is my decision, but the extinction of the Slender Billed curlew is a timely reminder that, at the very least, we could all make sure every shoot follows the advice of the GWCT recommendations for good practice when it comes to woodcock. I ask to go a step further because what 'grates my craw' is an uneducated shot killing a woodcock amongst a driven pheasant drive, with little or no understanding of how, or from where, that 'snipe of the woods' flew.

Shooting of woodcock - Game and Wildlife Conservation Trust (gwct.org.uk)

In summary:

1. AVOID SHOOTING WOODCOCK EARLY IN THE SEASON

Although migrant woodcock begin to arrive in October and November, at most sites across the UK, numbers continue to build throughout the autumn/winter and do not peak until December. Delaying shooting until many of these migrants, who originate from stable continental populations, have arrived reduces the risk of any possible impact on vulnerable resident populations. For those who shoot woodcock, beginning shooting after 1 December provides a helpful rule of thumb.

2. IMPROVE THEIR UNDERSTANDING OF THEIR LOCAL WOODCOCK POPULATIONS BEFORE SHOOTING
We advocate improving local knowledge about both the presence of resident breeders and the numbers of woodcock typically present at different times during the winter. Roding counts can be used to detect breeding woodcock where they occur and, over several years, provide an indication of local population trends. In autumn/ winter, a gradual increase in the frequency of woodcock sightings helps track the arrival of migrants and indicates the safest time to begin shooting (see above). This might be judged by counts of flighting woodcock at dusk, flush counts during autumn pheasant drives or seen at night during predator control activities. At most sites, migrants will vastly outnumber residents, providing a noticeable change in abundance once migrants have arrived. Information on how to count residents in summer can be found here.

3. SHOW RESTRAINT EVEN WHEN RESIDENT BIRDS ARE ABSENT
Restraint when shooting woodcock makes sense even in areas where there are no local breeders. We know from our satellite tracking and annual ringing of woodcock that the majority of migrant woodcock are extremely faithful to the same wintering site year on year. Overshooting will, therefore, break the migratory link with your shoot and is likely to lead to fewer woodcock being seen in the future.

4. SHOOT FLIGHT LINES WITH CAUTION
Shooting woodcock flighting from woodland at dusk carries a higher risk of overshooting. We advise gauging woodcock numbers across known flight lines first and then deciding on a bag, preferably only targeting half the flight lines once per season.

5. CURB SHOOTING IN FREEZING WEATHER
We have researched to understand the effect of cold weather on woodcock. Every effort should be made to reduce additional mortality when woodcock are at higher risk of starvation and predation during freezing spells. We know that most shoots stop shooting woodcock before a statutory cold weather suspension comes into force after 13 days. Our current advice is that shooters should stop shooting woodcock after four days of frozen conditions and allow the birds at least seven days to recover after the end of the cold period before shooting recommences.

JANUARY

In wildlife conservation, having a consistent and transparent domestic policy is essential to maintaining credibility and influence on the global stage. When a country implements and upholds strong conservation laws and practices, it sets a clear example that can inspire and pressure other nations to follow suit. This is particularly important in diplomatic negotiations concerning international wildlife agreements, such as those regarding migratory species and habitat protection.

BACK TO CURLEWS

With the **Slender-Billed Curlew** gone, we are left with our own **Eurasian Curlew**, still requiring significant help and in some cases as outlined above, better legal protection.

The hunting of Curlews is not restricted to Europe. In the US, one of the main historical threats to the Long-billed Curlew (pictured above) was hunting. Although illegal shooting still occurs, the main threats now include habitat loss and reduced prey due to pesticides.

Long billed curlew Illustration from ca 1830.

And before its presumed extinction, the Eskimo Curlew was one of the most common of waders in Northern America, with estimates of up to a million birds shot per annum.

Amongst the threats of human disturbance and habitat loss and degradation, the Far-Eastern Curlew still faces the threat of an estimated 2,000 shot per annum in Russia, in its breeding grounds out of a total population of 35,000.

And let us not forget the smaller version of our curlew – the Whimbrel, with its dark crown stripe, which is mainly seen on Spring migration in the UK, en route from Africa to its main breeding grounds stretching from Greenland to Siberia. It is a constant and joyous companion of mine when fishing in Iceland with its mellow whistle, a monotone pitch of exactly seven notes. A small breeding population exists in Shetland, where like the Eurasian curlew, numbers return to territory in April.

So we have lost another species

I am not suggesting the woodcock is threatened with extinction, but for those of us who care and are involved in shooting, we should take a moment to reflect and ask ourselves.

"Do we really need to still add woodcock to the bag on a shoot day"?

Wiltshire Countryman Sid Vincent

FEBRUARY

A WILTSHIRE MACNAB

THE FALLOW

It is 6:15am and still dark as we quietly close the doors on the pick-up. The first pigeons are cooing and somewhere high up in the sky a skylark is telling us that winter will soon be behind us.

We walk carefully up a rutted chalk track following an old stone wall. To our left is a wild, dense, beast of a wood filled with both ash and beech trees of all ages, weathered and gale broken; while to our right, pans a wide rolling expanse of winter barley. Soon, the fallow deer will return to this wood to lay up for the day - away from Sunday walkers, trail bikes and dogs.

All of a sudden, our stalking guide, Sid, freezes, looking straight ahead. Although the morning light has not quite chased away the night's darkness, we can just make out the even darker backs of our quarry (a term used in hunting to refer to the animal that we are pursuing). The deer must be approximately two hundred yards away - too far for us to shoot in the still dim, dawn light. You don't take chances with deer, and it is a 'golden rule' . Only take the shot if you are sure of a kill. The deer is still, you are a proficient shot within the range, and there is a clear view, for the intended passage of the bullet.

We continue with soft, steady steps. Heads down, lost in the shadow of the wall until we are around one hundred yards away. Carefully cradling the .243 calibre rifle in the mount of the shooting sticks, I push the safety from white through to red with a soft click, leaving my finger outstretched ready but not yet touching the trigger. My feet are planted securely as I move the cross hairs within the mix of dark brown heads, legs, necks and backs.

One deer is looking straight at me, I think we have been rumbled. The herd mills around seemingly uncertain as to where and what danger is upon them. One strides free, oblivious to its broadside presentation, stock still in the cool air. I move the sticks to the right, steady, and at first struggle with the light to find the mark. A little pale area of skin betrays the right shoulder.

Crosshairs on target, I move them a couple of inches higher and with a slight squeeze of the trigger, a distant thump disturbs the morning calm. Upon impact, the bullet makes a hollow sound as it enters the rib cage, finding the heart and lungs and instantly killing the deer before it then passes out on the other side. One hundred yards away, a large pale mushroom in the grass shows the shot was true.

Upon this disturbance, the herd wheels over the brow of the hill - out of sight in a slight gulley around to our right. We wait, scanning the direction of their travel, and, sure enough, the deer pop up over 200 yards away presenting a second opportunity.

This time around, the distance is more challenging, but it is lighter now and the crosshairs easily find their mark; 6:45am and we have two deer on the ground. It is now time for us to leave them in peace to their daytime schedule.

Hearing the rifle crack, a raven *"kronks"* that deeply powerful voice that splits the sky. Experience has taught him that rifle shots mean there is gralloch[5] to feast upon.

JOHN MACNAB: A TALE OF ADVENTURE AND REDEMPTION IN THE HIGHLANDS

I have not explained, until now, that we are after a Wiltshire Macnab (well, that is what we call it, being short on grouse and salmon) - to shoot in one day all three of the main species of deer that we have here in Wiltshire: a fallow, a roe deer and a muntjac. In saltwater fishing parlance, it could be a Flats Slam.

The original novel 'John Macnab' was written by Canadian writer John Buchan. A well-travelled politician, soldier, historian, poet, barrister, journalist, novelist and publisher, he was originally born in 1875 in Perth, Scotland. His book 'John Macnab' was first published in 1924 and begins with three gentlemen - a barrister, a cabinet minister and a banker -confessing to their 'ennui', a feeling of weariness and dissatisfaction with life, or what could be described as a sickness or malaise that no doctor can cure.

What the three men need, they decide, is a challenge: a devilish, daring challenge. And so, the drama unfolds as they send a letter to three highland estates, signed collectively as John Macnab. The letter informs each estate of their intention to poach a deer and a stag at a given time, before presenting the bounty on the door of their houses, upon which they rest their reputations. The modernday 'Macnab' was derived from the book by the Field Magazine and now involves a salmon, a deer and a brace of grouse all within one day.

---

[5] *The internal organs of a deer are ideally removed immediately after death to prevent the meat spoiling. This carcass is entering the food chain and must be treated accordingly.*

THE MUNTJAC

If it was a true Macnab, we would already have the salmon in the bag - the fallow being the hardest to find, stalk and then take down with a shot.

Up next for us today though is the muntjac, luckily it is a mild day, since 'munties' hunker down in the cold. Even so, we head for a sheltered wood with dark dens of drooping ivy among the yews, their lower branches neatly trimmed up to three feet off the ground by the little barking deer.

Two steps forward and we scan the terrain using binoculars, looking for the tell-tale chestnut coat standing out in the dark, the stripy face and gleaming marble eyes. A trotting sound to our left signals a muntjac saw us first and is off into the undergrowth with a brusque haughty bark to rub it in.

We walk on, it is full daylight now. Behind us to the north is the Oxfordshire Vale of the White Horse; to the right, the Wiltshire and Marlborough Downs; and directly ahead, due south, the approaching Lambourne Downs where the beautiful little chalk stream river rises in Berkshire.

Half a mile away, a herd of roughly 100 fallow deer are straddled out, their dark coats picking their way like cattle; but unlike cattle, they are wild and free to roam the three counties. Searching for a muntjac as the second quarry is like walked-up grouse shooting. We are creeping - gliding almost - eyes straining through the trees, watching the crackling dead branches with every step of our wellies. The muntjac has the upper hand with superb hearing and a keen sense of smell. But unlike the monarch of the glen; among all the woodland colours and with our binoculars we may just have the upper hand when it comes to vision. It is like finding a jigsaw piece and I have just found the colour orange. I bring the rifle scope up for a better look and yes, I can just make out a dark eye and facial stripe.

Carefully I set the sticks, cradle the rifle, find the eye again, and adjust the left leg so I am stable - but still, she only reveals her head. A pigeon clatters off and the deer's ears twitch, the game is up as he trots behind a yew tree. Quickly I move the sticks and have the gun aimed at a small clearing in the direction that he is going. His white tail briefly signals his trajectory. My eye goes back on the gunsights, my finger ready, hovering just above the trigger.

A walked-up grouse gives you a few seconds to raise the gun. The muntjac may just run across the opening, giving me no opportunity but in this instance, he stops. I have my chance and I take it. It is 10:30am; we have the fallow and the muntjac. Now for the roe deer.

THE ROE DEER

We drive west to the Marlborough Downs in Wiltshire to stalk a long belt of beech trees - in particular, the hedgerows and longstanding grass on the field margins where roe often like to lay up in the day.

There is a particular field I am looking forward to seeing, as there is often a flock of lapwings, redwings, and sometimes even golden plover to be found at this time of year. There is little bird life to be seen today, but sitting in the middle of the field are three roes, a buck, and two does. This will be a tricky stalk as they are at least 400 yards away and are all in the open. On top of this, there is no cover on any of the boundaries. However, on the far side there is a slight undulation which may allow us to crawl or walk to get in range.

Parking the truck on the far side of the field out of sight of the deer we reach a lone, windswept and bent hawthorn. The wind is good and the tree gives us a good marker. Sticks ready as we are on a slight brow that has shielded us from the deer's vision.

Steady paces, rifle ready. But, as we peer over the autumn stubble, the quarry is now out of range, walking away to our left. Whether you are John Macnab or John Miller, hunting opportunities come and go; but, fortunately, the deer are now making their way to a fence straight ahead of us, around 300 yards away. There is an opportunity after all.

Heads down, we walk steadily to within 200 yards of the fence as the first deer takes a leap over it. Sticks quickly up, rifle cradled ready for the second deer which is moving ready to jump, but I cannot get a clean shot. The third deer, ready to take its turn, is then stopped dead in its tracks with a thump. Finally, 12 noon for my Wiltshire Macnab.

A moment ago, these wonderful creatures were living free, roaming the land. My bounty from the day will provide a mix of choice cuts for two restaurants and a butcher, some sausages, as well as gifts for friends. And the amazing thing is that for all this meat, there has been no impact on the environment and the animals have not been couped up or been made to suffer a death in an abattoir: the deer being grass fed and having lived a natural life in the wild until their dying breath - their simple provenance, the fields all around me.

And after all this, I will still make it home in time to watch the Six Nations in the afternoon. It can be a blessed land we live in!

FEBRUARY

A CASE FOR STALKING

I only took up stalking when we moved to Wiltshire in 2012. I had always thought that deer were too big an animal to kill. However, I was introduced to Sid, a local ex-copper, who had shot and stalked all his life around Marlborough.

Sid is a countryman in the truest form, so every time we venture out together we discuss all the obscure sights and sounds we may observe – whether nests, frogspawn or birdsong.

When shooting, safety is paramount. Rifles are obviously dangerous, but Sid put me at ease right from the get-go, telling me gently but firmly that if I was not happy with my aim, if it was not a clean shot to make the kill, then I must not take the shot - it was always my decision and I had to use my instinct.

This is what stalking is all about. Is the shot safe? Check the background, will the bullet go to ground? A horizontal shot can go through a hedge and across a footpath - if you can kill a large deer (the size of a small cow) at 300 yards, then think what can happen to a person. You have to ask yourself whether you can clearly place the bullet in the area of the heart and lungs? Is the deer still, enabling you to take the shot, and are you comfortable, feet steady and rifle cradled securely?

As a man of the country, you respect your quarry and the thought of wounding it and causing unnecessary suffering makes you check everything. Experience helps you speed up these checks – hesitate and the opportunity may have passed.

You pick your animal and you pick your shot - take one out but leave the rest. We are not culling, we are harvesting one at a time with a single shot before moving on. We are part of the ecosystem, the apex predator doing what we do sympathetically with the environment. I think this is why I love it so much. You cannot say the same for all driven shoots or catching a salmon only to kiss it and let it go. Stalking connects you to your forbears all those centuries before you – there is a purity and a cycle to it.

SPRING COMETH

Splashes of white can be seen on the country lane as I run down past the West Overton church. In the trees above, a mighty cacophony of "caw caw" and constant bickering tells me that the rooks are reaching peak nest building. If they have a choice, they appear to favour the broader platforms of the large horse chestnut and sycamore trees, followed by oak. The beech and lime trees scattered up the lane are largely left ignored. On top of this, live nesting material is preferred over dead - presumably being more flexible to weave amongst the branches and its ability to bend with the wind; and perhaps the birds find it easier to snap off the more brittle chestnut and sycamore twigs.

The spring sound of "cuckoo cuckoo" is not yet upon us, but another seasonal sound is making itself heard as I approach a pond in Savernake Forest. Even before I can see the water, a different spring cacophony of croaking "rebbit rebbit" hits my senses.

FEBRUARY

Habitats are the cornerstone to flourishing wildlife. This pond is in the middle of the Savernake Forest and with no nearby roads and no surrounding agriculture, the result is a thriving mass of frogs every springtime.

And then three weeks later the same pond, all the frogs are replaced with toads making an almost Canada goose-like chorus – what a place.

And then three weeks later the same pond, all the frogs are replaced with toads making an almost Canada goose-like chorus – what a place.

Along with the frogs, the local birdsong is changing and the song thrush pushes his chest out to take centre stage in front of the mistle thrush, while towards the end of the month the gentler, sweet-sounding blackbird begins to shyly find his voice.

The wren and robin are also becoming increasingly loud, and, in the wings, the hedge sparrow begins to make his presence heard. So often overlooked, the hedge sparrow is never far from cover, skulking around the bushes, but as the days pass and start to become warmer and longer, he starts to overcome his inhibitions with sudden bursts of bravery. I am here, he is singing, let it be known.

Mild cloudy days may bring out the first "pink pink" of a chaffinch, the chatter of a goldfinch, and a high-in-the-sky skylark. All this is the beginning, the prelude; and the longer days give yet more hope for the spring to come.

RECIPE: SHEPHERD'S PIE (SERVES 4)

We are still at that time of year where we all want to be eating comfort food. For me, it has to be shepherd's pie – the perfect meal to eat on my lap watching the Six Nations rugby beside a blazing log fire. What could be better to see out the last days of winter? For a crispy shepherd's pie here is my recipe:

INGREDIENTS

500gm lean mince (<10%)

Two tbsp olive oil

Two leeks or four shallots

Three garlic cloves

Two jalapeno peppers

Ten gm butter for basting dish

Two tsp allspice

One kg potatoes (Maris piper, King Edward or Desiree) + butter

Sea salt and freshly ground pepper

INSTRUCTIONS

Number one rule.

No carrots or celery, they make it make it wet and stodgy.

First, fry some leeks – I recommend equal quantities of mince and leeks - in olive oil until you get a slight caramel edge to them.

At this point, add some crushed garlic, sweating it slowly with the heat down so as to avoid the garlic burning.

Next, add some finely chopped chillis and or ginger to taste.

Put your leeks, garlic, chili and ginger to one side, and, in the same pan, brown the lean mince (<10% fat) or minced steak. Make sure to drain off any excess fat at the end.

Next, mix the leeks and garlic with the mince while giving some good twists of pepper and salt. I also like to add at this point a good serving of all spice, chopped fresh rosemary and thyme (treat them as part of your five a day.)

Place your mince mixture in a metal pie dish basted in butter, then cover with good buttery mashed potatoes, and cook for just over an hour, until the top and sides are crispy.

Serve with frozen peas or baked beans if you are a teenager.

FEBRUARY

I asked David White, a superb wildlife photographer in Wiltshire, if he could get a picture of a robin on one of my spades. The robins were shy, so instead, here is a wonderful photo of blue-tit in action.

GETTING STARTED: THE KITCHEN GARDEN

In mid-February, I sow the first seeds of the year in a heated propagator inside the greenhouse: chillies, tomatoes and summer bedding.

Each year, I experiment with all three crops, and this year I have decided that a medium jalapeno will stay in the greenhouse all summer long, while I will plant my Padrón pepper outside later in May. For the tomatoes I have gone for a large ribbed Costuluto Genovese for cooking and for use in salads; plus, a Dr. Carolyn Pink for pure flavour. And finally, a mixed colour, tall antirrhinum (snapdragon) that will flower until the first frosts and is so loved by bumble bees.

Chillies, in particular, take a while to germinate and are slow to grow, so it is important to start them early in the year. The alternative is to buy seedlings; but sowing my own gives me the opportunity to keep trying new varieties and also provides some activity in February while most of the garden sleeps. The chillies are the first shoots of spring in the Kitchen Garden, and like a first born, get more than sufficient attention.

If the forecast is not too cold, I also start potting up my dahlia tubers, releasing them from their winter shelter of my hessian sacks (even within the greenhouse I will keep the sacks handy to throw over the pots in case colder weather reappears). For me, the dahlias are a symbol of the seventies, much like Abba and Slade. I can remember these bright and opulent flowers growing in the front gardens of the houses in Crowsley Way and Kidmore End, but then for several decades became outlawed, considered as naff by 'snooty' gardeners. In recent years though they have made a big return, probably in part thanks to both Sarah Raven and Monty Don who have sung their praises. To me, dahlias are a bit of fun, an explosion of colour and texture. Certainly not 'naff' but definitely a bit of a 'faff' to look after.

Not only is it sustainable, but there is a traditional charm to digging over the vegetable beds, covering with home-grown compost and then cutting my own hazel for pea sticks and bean poles. They may only be bare sticks today, but once pushed into the soil their neat rows are a harbinger of a Kitchen Garden soon to awaken.

We only see siskins in the garden towards the end of February before they disappear again by March.

David White set up a hide to take this picture in our front garden.

MARCH
THE DAWN CHORUS

There have been some interesting articles recently questioning whether we are doing more harm than good by feeding our wild birds. The arguments persist due to a build-up of unnaturally high densities of birds in relatively small areas. Well-intentioned it may be, but the British public is potentially disrupting a finely balanced ecosystem, resulting in the reorganisation and redistribution of different bird species across the UK. There are four main reasons for this disruption:

1. Increased disease transmission.

2. Increased predation:

    a. Direct predation, such as sparrowhawks and;

    b. Indirect predation through the attracting and supplementing of populations of predators who may prey on nests and the young birds later in the year, such as rats, corvids and squirrels.

3. An unnatural dependency that becomes critical when the food runs out;

4. Habitual feeders, such as blue tits, see their numbers boosted at the expense of more 'subordinate' birds, such as willow tits, who lose out as there is greater competition for nesting sites.

To mitigate this consequence of the additional nutrition we provide, I try to abide by certain rules:

a. Regularly cleaning the bird table and feeders to mitigate disease transmission, as well as limiting the time that the bird feed is in one place.

b. Keeping the food above 'cat' height, placing it near to bushes and shelter and making it as squirrel and rat-proof as possible.

c. Limiting the overall time I spend feeding the birds. Most wild birds should be able to fend for themselves until at least Christmas so concentrate on beginning sometime in January, dependent on the weather, and finishing around mid-March. Again weather dependent.

By providing food stations in winter, it is also incumbent on us to provide suitable, welcoming nesting habitats. Dense hedges and nest boxes provide ideal nesting sites, but what is often overlooked is the requirement for somewhere quiet and overgrown, where the newly fledged young can continue to receive support from their parents.

I have one such undisturbed corner in my garden with some yew hedging, tall mahonia bushes, dense holly, and piles of leaves and long grass. Every year, song thrushes, robins, blackbirds and dunnocks nest here. However, it is not just the immediate nesting site that attracts them; it is also the surrounding habitat that provides cover and protection for their young as they grow and strengthen their flight feathers during that critical period upon leaving their nest for the first time. By cultivating such a habitat in your garden, you're not only encouraging nesting but also providing a safe haven for fledglings within which they can thrive. In March, the nettles and cow parsley are now creeping up the stems of the bushes to boost this hideaway.

One of the birds that I occasionally spy on my bird feeders is the beautiful long-tailed tit. However, I have come to learn that their nests invariably end up being predated by magpies and so as I mentioned in the first chapter, I am keen to create a more suitable habitat for these little birds in my garden in order to give them a better chance. Long-tailed tits like to build their nests in impenetrable spikey bushes to help protect their nests and chicks - under the cover of some mature trees. I am impatient to see if the three gorse bushes which are truly spikey, along with two roses and a holly bush will attract them this year.

In the garden, two new additions have made their homes in a dying Sycamore tree. Rather than fell the tree to the ground, I decided to leave eight meters to allow nature to feed off its decaying embers – so it is rewarding to see this green woodpecker and his buddy, a great spotted woodpecker, chiselling out their nesting holes in this tree. In particular, green woodpeckers seek out dead and dying trees so it is satisfying to witness one here already. Jackdaws and stock doves also squabble over a split on one side of the tree, and I am lucky it is within easy sight of my office. Perhaps one day a pair of owls will move in.

NATURE REAWAKENING

This year, the river Kennet springs popped back in early January, but with little subsequent rain in February it is now clear and beautiful with the streaming brook ranunculus starting to emerge from the previously dry riverbed.

Crossing the bridge at West Overton - where it had been bone dry only 2 months ago – I spot a 2lb brown trout riding the current. A wonderful sight! That fish had travelled upstream, and my thoughts immediately turned to whether it would successfully navigate the return journey when the waters recede - as they inevitably do - back into the chalk come June and July.

The importance of the ranunculus weed on the stream beds all along the channel here cannot be overstated. Within a month it is choking the stream, raising the water level and, crucially, holding back the inevitable effect of gravity. Just below the bridge, there is a small lake which will also be dry come summer, but today and for now, it is much loved by mute swans, Canada geese, mallards, moorhens, gadwall and teal. Think what it could sustain if it flowed all year.

Back at home, a small bird caught my eye as something unusual among the common garden birds. And there, sure enough, through the binoculars was a green fluttering bird - bigger than a goldcrest, but smaller than a greenfinch - the first chiffchaff of the year. And then an even more delightful surprise as I raked over some old compost: the hidden gem of a perfect miniature grass snake. One of last year's brood. Only five inches long but with all the adult markings and that wonderful yellow collar. It was sluggish and lethargic in the March air so I carefully covered it with dry leaves and some twigs to stop any eyes in the sky from spotting a meal. A sure sign that the snakes had successfully nested last year.

The dawn chorus is thoroughly underway by mid-March, with song thrush, blackbird, robin, hedge sparrow, starling and chaffinch filling the garden with their song; while in the fields the "little bit of bread and no cheese" and the "jingle jangle" of the yellowhammer and the corn bunting accompany your walks along the lanes and bridleways.

Steeped in folklore, these 'elf cup' fungi can be found in late January if lucky.

LITTLE BROWN BIRDS AND POLYANDRY

Let's hear it for the little hedge sparrow. Also known as the hedge accentor or dunnock, which is derived from the old English word for brown 'dun' and 'ock' for small - so the 'little brown' bird.

This little brown bird quietly busies itself on the lawn close to the hedges and any shelter, poking furtively among your flower beds - afraid of its own shadow. You would hardly know it was there and perhaps you don't realise it is, in fact, there. But listen carefully in March and you will hear a beautiful sudden song from the top of a bush. Squeaky piping notes that are pushed out with purpose, quickly and briefly - endearing in their simplicity, as well as amazement at their volume for such a small bird.

I always point these little birds out to friends and family to give them some recognition. Viewed up close, their forms reveal a back with irregular brown stripes, a sleeker daintier bird than the house sparrow. The head is a dusty grey-blue centered with a bright chestnut eye, and perhaps that eye betrays the rather 'quirky' side of this little brown bird.

An incongruity that quite defies its creeping demeanour, as this bird practices polyandry – an 'arrangement' whereby the female can have several mates. However, the practice is rather unusual as the male performs a ritual before mating, where he uses his beak (with the females' consent) to remove any sperm in her cloaca from any previous mates. The male then copulates with the female himself, thereby hoping he is passing on his genes rather than his predecessors'. Never judge a book by its cover, or a bird by its plumage. Such a quiet undemonstrative bird.

MARCH IN THE KITCHEN GARDEN

There are many gardeners who say the gardening calendar starts in March but be warned that the ground may still be cool and frosts may yet still make an appearance. Charles Dickens made the point perfectly in 'Great Expectations' when he wrote:

> "It was one of those March days
>
> When the sun shines hot
>
> And the wind blows cold;
>
> When it is summer in the light
>
> And winter in the shade."

That said, even with a sharp wind, as the sun climbs higher there is an undercurrent that mother nature is building and growing - the exhilaration of spring, the promise of growth and abundance.

With these thoughts, the season finally starts in the Kitchen Garden with the planting of potatoes and the sowing of broad beans, instilling my connection to the seasons. These two crops are the embodiment of 'homegrown = superior flavour'. The potatoes I am growing are early salad potatoes, not main crop. We all have our favourites, and our geology and weather will determine what suits our garden, but the best variety for me is Charlotte.

With potatoes, there is a constant argument over whether you should 'chit' them for a few weeks before planting and whether there is any point. Chitting is the growing of little shoots that come out of the potato 'eyes' designed to give them a head start before they are planted. The simple answer to the argument is if you want the earliest potatoes then yes chit, as you can get them going while there are still frosts outside, but for maincrop there is no point. My soil is too thick with clay for a no-dig approach and part of the joy of starting my planting now is the digging of a small trench, halffilled with barrowloads of compost, and the placing of the chitted potatoes in their cosy bed, nestled about 18 inches apart with the shoots facing upwards.

During World War Two, the head of the local Home Guard lived in our current house together with his wife, who was then the local head of the Women's Institute. The couple wrote seasonal poems about what was happening in the village, one of which I have included below:

"Now it's potato planting time

Sew onions by the score

Of aching backs, hitching up slacks

Gumboots, that Father wore

Dig Dig Dig

Then dig and plant, don't say you can't

We must win this war

Just think it all depends on you

The million things that women do."

MARCH

Potatoes nicely 'chitted' ready to plant out.

Coincidently, I am writing this chapter on International Women's Day (8th of March) and the end of World War II saw a sea change in understanding and recognising women's contributions and standings in society, from the traditional and unoriginal troupe of child-rearing and housekeeping. You will also see in the poem words not commonly used today such as slacks, score, and gum boots!

My father wore gum boots, which were not tall like our wellington boots of today, and always with the top rolled over – why was that? And my mother wore slacks!

In the Kitchen Garden, I am pleased to see a few weed seedlings, meaning that the soil is now warm enough to sow my four rows of broad beans and plant out the red onions. Touch the soil to gauge whether it is too early to sow your seeds: if it is cold to touch then you need to wait a week or two more as you are looking to feel a neutral temperature - not warm but not cold either, with a perfect mix of moisture easily broken down into a fine sowing tilth.

Broad beans in the shops rarely have any flavour and these beans that I am sowing will be a treat from late June onwards. Covering them carefully and standing back to admire my handiwork a male brimstone butterfly floats across the garden – with its rich creamy yellow wings I imagine this must be where the word 'butterfly' came from. This delicate, sunshine-like little insect is a messenger telling me that winter is drawing to a close and that spring and summer will soon be upon us.

JOHN MILLER

EARLY 'FORCED' VEGETABLES: RHUBARB AND SEA KALE

Forcing rhubarb accentuates the flavour and colours.

MARCH

Chopped up it is like candy in a bowl.

'Forcing' rhubarb is a technique that is used to produce sweeter stems earlier in the season. By covering the rhubarb plant with a pot or bucket, or by bringing them indoors and keeping the in the dark, you can force the plant into putting more energy into seeking out the light, which results in the plant growing extra quickly, producing beautifully sweet, tall, pink stems.

In my garden, the first rhubarb 'forced' under a clay pot is now ready for picking. I say 'pick' but you actually 'pull' rhubarb so that it snaps at the base. You never cut with a knife or it will rot back into the base. I have found that by 'forcing' my rhubarb, the plant produces an array of incredibly bright scarlet and fuschia pink stems that cut perfectly – the stems are crisp with an unusual acidic smell.

Normally in March I boil the rhubarb with some sugar or honey and lemon accompanied with vanilla ice cream, natural yogurt, or custard. But then I quickly tire of it, so this year I am making rhubarb vodka based off the same recipe as the sloe gin.

Fill a sealed Kilner, with 60% vodka, 30% cut rhubarb and 10% caster sugar. Shake every day for about two weeks and then leave for three or 4 months before decanting, ready to drink.

To date, the rhubarb vodka has a wonderful pale Provence rosé colour which, I expect, will deepen as I leave it until the end of the summer. The question then will be where will I drink it?

Perhaps on a glorious Indian summer's evening or whilst waiting in a duck hide for the whistling wings of gadwall, mallard and tufties? Rhubarb is a naturally tough plant, being one of only a handful of perennial vegetables surviving in Finland, Iceland, and even Siberia where it is found in their traditional dishes. However, 90% of the world's commercial rhubarb is grown in Yorkshire, where it thrives in the cold, dark and wet winters.

The other 'forced' vegetable that I grow is sea kale ('crambe maritime'), which, unlike rhubarb, is more akin to asparagus with a subtle nutty flavour. It is a native of European seashores where its deep tuberous roots burrow through the shingle searching for water, surviving where few other plants can. The Victorian traditional clay pots in the picture are noticeably shorter than the taller rhubarb pots. I find that the stems are best steamed and then served with a little butter, salt and pepper. Sea kale is a good source of vitamin C, hence its traditional medicinal use for preventing scurvy.

MARCH

My sea kale pots in the spring sunshine

Having a pint outside a London pub after the book launch for Tarquin Millington-Drake's 'Living with Greys'.

EPILOGUE

Tarquin Millington Drake spoke passionately about how everyone who loves the countryside should give something back, a point that resonated around the room.

We all have our favourite animals, for Tarquin, this is the grey partridge. He spent five years photographing this elusive bird that is so full of character and fell in love with it. All beautifully explained through the seasons, with 200 beautiful photographs chosen from 45,000 images taken while sitting in a hide for around 500 hours.

This is the same man who brought us all together on the North Pennines trip, which I discussed in April, with an organisation he founded called Why Moorlands Matter (WMM). The aim of this organisation is to bring people together from different walks of life to have an open, honest discussion about what is right and wrong within the shooting community and how it can benefit conservation.

A grey partridge in the North Pennines. Photo by Tarquin Millington-Drake #whymoorlandsmatter

RECONNECTING WITH NATURE

The twin challenges of climate change and a reduction in biodiversity are real and scary. But they also bring hope and opportunity, with today's better awareness and modern solutions to tackle these related ecological challenges.

There have never been so many well-meaning and equally well-funded conservation bodies in the UK, and corporates bending over backwards to sink millions into 'green' policies. Yet we remain challenged and frustrated. At best we see a poor return on investment and, at worst, a situational irony with harmful conservation measures.

As I see it, there are two points that need to be addressed if we are to solve the challenges facing our natural environment:

> **POINT ONE** – There can be a cultural barrier between conservationists and those who manage the land. A prevailing 'us and them' mentality that holds people back, with both 'sides' becoming polarised and defensive - when what we want is a common way forward.

> **POINT TWO** – The second challenge, partly explains the first. A basic lack of knowledge of natural history causes those people on the ground (who may have a lifetime of experience working the land) to feel threatened and under siege by people without the fundamental working knowledge to make the call. And like all good arguments, there will be minorities at the extreme ends of both parties who are just deaf to change.

Each one of us is part of Mother Nature, part of an ecosystem we inhabit together, yet we have disinherited ourselves and jumped ship for quick fixes online and the ever hungry strive for GDP and growth. This decoupling with nature will only widen with the next generation of school leavers and be compounded by the deaths of an older generation whose lives were more physically bound to the land.

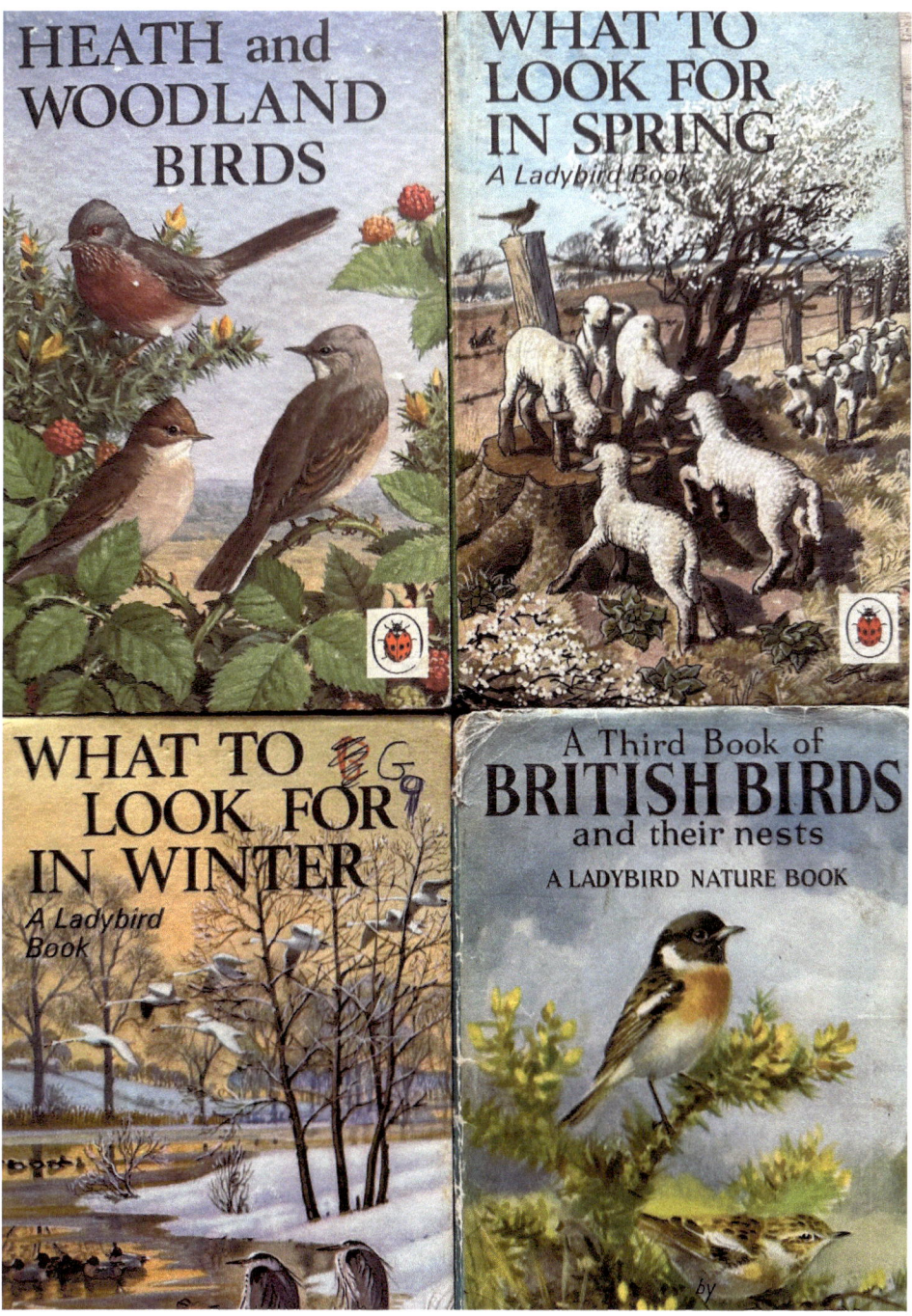

I was inspired by nature as a child by the simple but beautifully illustrated Ladybird books.

That said, it's no use complaining and pointing fingers as change starts with planning and preparation, and the ultimate change agent is education. In this vein, I have been repeatedly asking myself, have we been missing something here?

The new GCSE for Natural History is a first step in reattaching this link between man and nature and addresses the aforementioned point two. Society needs naturalists more than ever before to observe, monitor and record the natural world and understand the changes taking place. These are essential skills for the emerging green economy, in both forming and adapting new environmental legislation.

It may also save the larger NGOs from becoming a dysfunctional mirror of the public, by shifting the accumulated knowledge from pushed content to real life practical experiences. Note, a PWC report in 2023, 'Dependent on Nature' suggests that 55% of the world's GDP is reliant on nature. Our lives are completely intersected with nature and we would do well to remember that when nature thrives, so too do humans.

It should not be a surprise to know that our children have become disconnected from nature. While our urban population has swelled by over 15 million people since 1960, the rural population has shrunk by nearly a million. And on top of this, nature as a formal part of a child's education is completely lacking; and even with the best intentions, it cannot be mitigated by a few brave and passionate teachers and naturalists.

Add to this the invention of video consoles, more houses and smaller gardens, fragmented families and social media (through which so many people seem to live virtually), who is there to fire up the imagination of a child's natural curiosity? What happens when my generation disappear - who is left to raise the standard?

It is no wonder that it is hard to argue on behalf of the country as half the time no-one really understands what we are talking about. For example, people, lots of people, really do continue to eat open-net farmed salmon. Maybe if someone had fostered in them as a child a love of nature they would be a little bit more curious and would feel more empowered to use their purchasing power to send a strong message to the salmon farmers: "Your salmon farms have awful husbandry and present a terrible threat to the natural environment – on top of this, your business models are unsustainable. It is time to stop!". The cruelty alone is another reason to ban open-net farmed salmon - I recently read that nearly 15 million salmon mortalities were reported in fish farms across Scotland between January and November of this year by the Fish Health Inspectorate (FHI). Imagine if that was cows dying in the fields – the consumers and farmers would be in uproar.

We have turned our backs on the legacies of Charles Darwin and the Mary Annings of this world (wonderfully described as naturalists, explorers, geographers, anthropologists, biologists and illustrators). And with them, we have also turned our back on wildlife. Both to our peril. True, we are very lucky to have the mystical David Attenborough, but he is often whispering his magic in some far and foreign land. As a result, our children may know more about the plight of dwindling aardvarks in the Kalahari than the sticklebacks in their local brook.

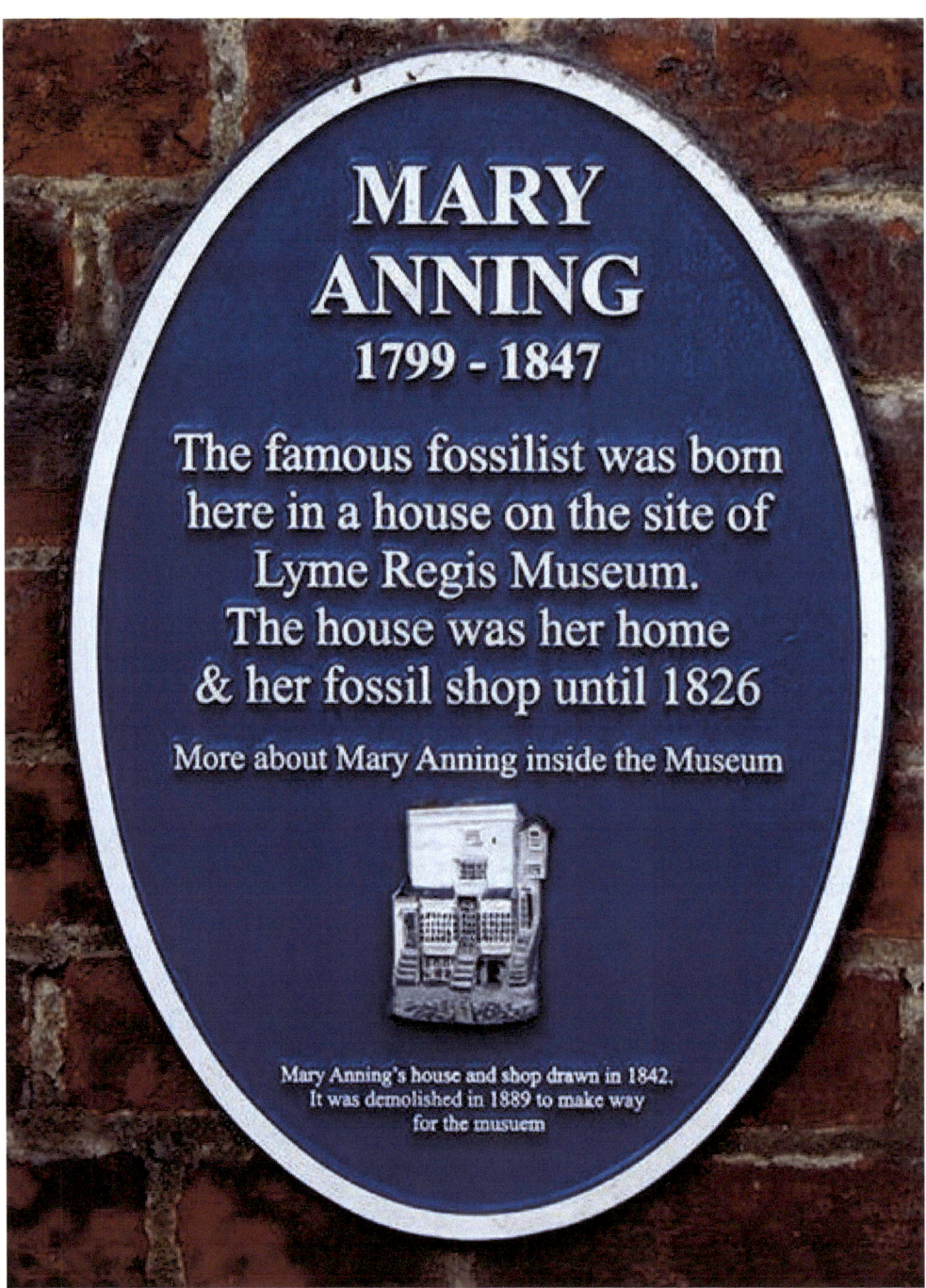

We have many famous naturalists in the UK.

Our children will grow up believing that our impoverished world with fewer species, less biodiversity and the huge reductions in absolute numbers - the bio abundance - is the accustomed, habitual standard. As a result, a new low bar is drawn in the sand. We have all become incapable of connecting the dots at a time when the environment is crying out for people who understand the bigger picture.

It is interesting to learn that a reduced awareness of nature during a child's formative years, as well as a lack of appreciation for the natural world around us, can also generate apathy for the very environment we have to protect. We need nature-literate students in big business and industries such as energy, construction, agriculture and fisheries, as well as the public sector, to create and hit our environmental and climate goals.

Over the last 20 years, we have found room for much diversity, which is a good thing. But crucially, we also need to find room now for biodiversity.

Governments, businesses and individuals are all so fixated on GDP as the cure-all measure of life that we lose the excitement of the everyday wonders all around us. Most critically, we also lose that sense of identity and a sense of connection to our local wildlife and plants. Our local flora and fauna belongs to our community - it is our local habitat to steward and care for, to nurture. Having a link to your local wildlife and plants can create a spiritual connection to a place, a sense of belonging, as well as a sense of peace.

But today, as I draw my year's thoughts to an end, I want to draw your attention to an opportunity we have to reverse this nature-deficient trend in the UK and reverse the nation's widespread lack of understanding of the natural world.

Nature has long been sidelined, we have forgotten how to live hand-in-hand with her, constantly expecting her to bend to our will and our desires. However, this new 'Natural History' GCSE has been created with the goal of creating and nurturing generations of nature-literate individuals who will be able to positively influence the national debate. A pivotal moment as new GCSE subjects are not regularly introduced to the system at all. The benefits to come from this new Natural History GCSE will include:

> **EMPLOYMENT:** This qualification will help individuals to be better prepared for the array of new jobs in the green economy. A core foundational grounding in nature will bring greater empowerment to employees and better qualifications for employers. This newly attained understanding of nature will not only positively impact those working directly with nature, such as farmers, but

will also help to ensure that businesses across the spectrum recognise the importance of nature and as a result ensure a 'Natural Capital' approach.

EDUCATION: At the moment, we lose children and teenager's curiosity in the natural world as there is a gap in the curriculum. Students could experience their complete secondary education with no formal education on the natural world, except during some geography lessons perhaps. This new GCSE will spark the interest and passion of students whilst they are still young; and as it is a formal qualification, it will enable natural studies to be taken seriously as a subject.

There is a wonderful opportunity to not only connect children back to nature but to also give them a sense of a local and community identity with their British flora and fauna.

Finally, you could also argue that the biodiversity crisis is being overshadowed by the climate crisis. Without understanding biodiversity, we too often make uninformed decisions without understanding the picture as a whole. Equally, if students understand biodiversity, they may feel more empowered to step into the climate debate.

HEALTH: There are many well-documented papers explaining how a more nature-literate society can lead to improved mental health. It was only during covid and its ensuing lockdown, that many people finally began to appreciate the role nature can play.

SO, WHAT EXACTLY IS NATURAL HISTORY?

The Natural History GCSE focuses on understanding the natural world in all its richness and diversity. Through observational study and investigation, Natural History seeks to understand the complexities and interconnectedness of life on Earth in all its contrasting habitats. The GCSE explores how our natural world has been shaped and how it continues to change, both by natural processes and through human intervention.

> *"The GCSE in Natural History would reconnect our young people with the natural world around them. Not just because it's fascinating, not just because it has benefits for mental health, but because we will need these young people to create a world that we can all live in - a vibrant and healthy planet."*
>
> **MARY COLWELL, AUTHOR AND ENVIRONMENTALIST**

HOW DOES THE DEPARTMENT FOR EDUCATION DEFINE THIS QUALIFICATION?

This rigorous qualification will allow the study of the natural world in a way that other GCSEs do not – considering all aspects in depth, as well as assessing the wider context. This GCSE will complement and be distinct to other qualifications that consider aspects of the natural world, such as Geography and Biology.

A GCSE in Natural History will develop students' skills in observation, monitoring, analysis and reporting; in turn, this will support students to develop a deeper understanding of the complexity of the natural world, including its fragility and interdependencies. The Natural History GCSE will also involve active outdoor engagement - whether they live and study in urban or rural environments - including assessing students' local wildlife and ecosystems.

As part of their study, students will not simply learn general definitions of habitats and ecosystems, but they will look at specific habitats, including the flora and fauna, and the inter-relationships between specific examples. Students will also explore the natural world at a local level so they can name, classify and understand organisms – they do not merely look at ecosystems and processes, such as pollution, as case studies; instead, it is a richer, more indepth learning experience, underpinned by engagement with and observation of the natural world by:

- Encouraging detailed study of **specific organisms (wildlife) and the environments** in which they live.

- Developing a student's knowledge and understanding of the forms, functions, and behaviours of wildlife through the study of **real plants and animals** across a range of settings.

- Developing a student's appreciation and understanding of **critical interdependencies and relationships with other species**, and between specific organisms and their environment - locally, nationally, regionally and internationally.

- Helping equip young people with the knowledge to help tackle and better understand future **environmental challenges.**

Photo credit Action for River Kennet – ARK

The introduction of this new GCSE will help formalise nature within the curriculum, empowering people to become nature literate and providing a pathway to further education in nature related studies, as well as a nature-positive economy.

And we the parents have a part to play – it is something to talk about in your next parent's evening.

Whilst on the topic of school and education, it is worth noting that for some time, there has been a growing debate in the education sector about the right balance of coursework versus exams. Of course, the answer varies by child and subject. However, within this argument, there is also a discussion of how a learning pathway with real-life experiences is more effective for some students than testing everything with a final exam. The Natural History GCSE can help here, providing an example of alternative learning and testing. Unfortunately as we go to print the new Government has delayed further the introduction of this GCSE

Finally, if we want our children to have the fun and curiosity that I experienced throughout my childhood and which has taught me so much, then it is up to us as parents to open our children's eyes and instil a love of nature. We must also educate our children so that they not only love, but also understand nature and will fight to cherish and look after it for everyone. I began this book asking people to look and see everything around them in the natural world.

I end the chapters asking all parents to help their children to look around them and be curious.

There is always something to see, something to learn from, and something to nurture. We must simply open our eyes and hearts to our surroundings.

PHOTOGRAPHS

Charcoal and crayon sketch of myself as a boy circa 1971 by Tom Coates (1941-2023) PPNEAC, PPRBA, PPPS, RWS, RP

APRIL

Mrs. Owl keeps watch from an old Scots pine tree. 10

The baby owl in her crib. 12

Out of the crib and growing strong in one of our Rowan trees. 13

It was not until May 8th that Mrs P. finally sat down to incubate her 15 eggs. 16

A male grouse: one of the UK's most wonderful birds, yet its survival is threatened. Photo by Ian Harrison. 19

Among British birds, only the even rarer capercaillie perform these leks on traditional open spaces year after year. Photo by Ian Harrison. 20

Curlew in the North Pennines with its charismatic long curved bill. Photo by Emily-Graham-media. 24

After two or three casts; ice begets ice, and you have to break the build up from your rod rings. 26

One of the joys of fishing is coming close to these beauties – a wonderful cock sea trout. 28

Top: Under the lime trees in our front garden is the perfect habitat for these wood anemones. **Bottom:** By April, the brook ranunculus are flowering at the head of the Kennet, long before its bigger cousin downstream which requires water all year round. 30

Diagram reproduced by kind permission of Atlantic Salmon Trust. 31

These two photos, taken on the Upper River Dee, show the difference that habitat restoration can make, with great progress currently underway. Photo credit: River Dee Trust. 32

Diagram kind permission of the Atlantic Salmon Trust. 33

MAY

Mechanical sluice on the river and it still works! 36

When the Kennet trout throw themselves out of the water. Photo: kind permission by David White. 40

West Kennet long barrow on a May morning, looking South towards Devizes. 42

This is one of my favourite nests within which I found the long-tailed tit. One of their favourite habitats is this spiny red leaved 'Crown of Thorns' bush. 44

Top: Is there a friendlier summer sound than the constant chatter of house martins? **Bottom:** …or a more adorable sight than a newly fledged goldcrest. 45

Top: Diagram: With kind permission by Atlantic Salmon Trust. **Bottom:** The rotary screw trap in operation. 46

Note how this salmon smolt has 'silvered up' from its original golden brown parr stage before it's migration to the sea. 47

Recipe: Watercress Soup. 50

A heavy shower emphasises the beauty in the fresh growth of water lillies and marsh marigolds at the end of May. 53

JUNE

Looking across the River Dee at Carlogie to the Kincardine hut on the opposite bank. 54

The Blue Crathie, an excellent low water fly on the Dee. The blurred white building behind is Dess Mill. 56

Top: *We are very strict these days on catch and release but I am not sure what the rules were back in 2012 for kissing your capture. This beautiful salmon was 24lbs and caught with a size 14 crathie, mid-afternoon on the 4th August 2012 at Middle Gannets Ballogie.* **Bottom:** *A heavy night after fishing for sea trout. 58*

Top: *The Carlogie Hut is a place of dreams and tales and roasted Wiltshire muntjac!* **Bottom:** *Prepping for the trip is half the fun! 59*

Recipe: Seasonal Frittata. 62

Top: *Seasonal frittata.* **Bottom:** *Garlic scapes (flower heads). 65*

Top: *Let part of your lawn grow wild, the bees and insects will thank you (and in turn, we will thank them for their integral role they play increasing biodiversity and our crops).* **Bottom:** *I wonder if spotted flycatchers were originally named after their young in the nest, as the parents are distinctly streaked. Here they are nesting on one of the purlins protruding out of our house. 68*

Top: *Snapdragons, linaria and foxgloves reach a crescendo in late June.* **Bottom:** *Yellow rattle, the keystone species to a successful wildflower meadow by weakening the more dominant grass. 69*

I will always love this simple, small and elegant wild dog rose over any cultivated variety. 71

JULY

Showtime for the greenhouse. 72

Looking downstream from the pumphouse on the river Kennet at Eastridge. 75

Recipe: Cirtus Celeriac, Fennel and Redcurrant Salad. 76

A beautiful 7lb leopard spotted Testwood sea trout. 80

Top: *Willowherb is a great foodplant for elephant hawk-moth caterpillars.* **Bottom:** *The aptly named fox and cubs (Pilosella aurantiaca) where the buds are the cubs and the open orange flower the vixen. 81*

AUGUST

Note the tiny holes in the tube of these hitch flies which skate across the surface. 85

River Halle, Skagastrond. 86

Top: *Manfoss waterfall Laxa Asum, no barrier to the Atlantic salmon.* **Bottom:** *A sheltered grassy knoll is welcome after 10,000 casts. 87*

Recipe: Runner Beans & French Beans Done Three Ways. 88

Recipe: Garlic, Chili, Lemon & Thyme Spatchcock Chicken on the BBQ. 90

By doing the "Chelsea chop" (where I cut some of my herbaceous perennials back by a third in May) I can extend their flowering season by a good six weeks. 93

SEPTEMBER

The Alladale Estate in late August – you can see with the regeneration of mixed woodland. 94

Top: *The wind comes straight off the North Atlantic, so we hide in here with a whisky until the wind dies down and we are ready to cast our lines again.* **Bottom:** *The Laxa Hrutafordur is shaped by natural larval dykes cutting across the river's path, creating a series of waterfalls and deep pools. 98*

Teddie stripping salmon eggs to help spread them across a wider catchment. Note the ice along the river. 100

The adult salmon can only survive in the tiny pots of deeper water at the bottom of these waterfalls, waiting for September rains. 102

A typical Hrutafordur September salmon. 103

Curlew North Pennines 'back on the moor', photo by Emily-Graham-media. 107

Ground nesting and in the open they struggle with predation of both eggs and chicks. Photo by Emily-Graham-media. 108

Curlews usually lay a clutch of 4 eggs, hidden in a tussock. 109

Curlew North Pennines 'back on the moor', photo by Emily-Graham-media. 114

OCTOBER

Loch Horrisey photo courtesy Jonathan Evans. 116

Valley House, now a ruin, has remained untouched since the 1920s following the tragic death of Beveridge's son. Photo courtesy of James Macletchie. ©www.jmacletchiephotography and www.igot2travel.com 120

When it is only 6C, the fish are nowhere to be seen or hooked, and the sun makes an appearance - what do you do? Find a sheltered nook on a grassy knoll, drink whisky, smoke cigars and talk total bollocks. It is almost as enjoyable as fishing! 121

Top: *Changing light North Uist sunset looking West 17:24.* **Bottom:** *Changing light North Uist sunset looking West 17:43. 123*

Top: *A Scottish farmed salmon with severe sea lice damage – unfortunately this is not an isolated incident. ©CorinSmith.* **Bottom:** *Nearly half of Scotland's salmon farms burn, dump or destroy millions of dead fish every year, according to Scottish Government data analysed by The Ferret. ©CorinSmith. 126*

Red campion and cow parsley a magical mix. 127

The village of Avebury in Wiltshire. Photo courtesy of David White Wildlife. 130-131

Top: *Garden produce in October in our larder at Gypsy Furlong.* **Bottom:** *Padron peppers cooking on the brazier. 133*

Recipe: Roast Celeriac. 136

NOVEMBER

The Viola Tricolor, a field pansy also known as Love-in-Idleness and Heartsease. 138

The English television presenter and writer, Jack Hargreaves OBE – a true countryman. He dedicated much of his life to highlighting the accelerating distortions in relations between the city and the countryside. 143

Elephant garlic (the large bulbs) and Caulk Wight drying in the sun after harvesting. 144

Fresh lemon sole and squid from Lyme bay Dorset English Channel. 147

Recipe: Bone Broth. 148

A Burst of Colour. 152

Remaking the Wild – It's Personal. 153

The same smolts by November. Photo: With kind permission from Atlantic Salmon Trust. 155

Top: Large woody structure (LWS) on the river Muick. Photo: With kind permission from River Dee Trust. **Bottom:** And here is the same structure back in the summer, showing salmon using the woody structure. Photo: With kind permission from River Dee Trust. 156

Looking downstream from the bridge where the glide drops into the fast water. 157

Peering into another world. 158

DECEMBER

Cigar Club Winter Quarters where we can plan next year's adventures. 160

Mrs. Owl got used to me and just sat looking down with her big round eyes – I think I am in love. 164

The Longhorn beef is mind-blowing. 166

Stork nests in the oak trees at Knepp – perhaps this is a throwback to their preference before modern buildings and latter-day cartwheels? 167

Finally, Dominika comes fishing although this is more like a fishing heaven – Alphonse in the Seychelles Outer Islands. 169

The stripey torpedo-shaped bonefish with its spectacular triangular, translucent fins and tail. I think the best way to describe it is a saltwater barbel. Add the fact saltwater fish are generally more powerful and you get a feel for its incredible strength in bending your rod. 170

Giant Trevally. Gary said "lift it higher, I said I am trying!" 173

Bluefin trevally caught on a surf walk, being pounded by waves casting into the breakers. 174

It was an honour to hold this beautiful milkfish and then watch her massive tail with a sharklike wiggle propel her out of sight into the deep. 177

Big eyes, big mouth but no teeth! The milkfish is a gentle giant. 178

Eurasian hobby wintering on Alphonse Island. 180

Even in December there is joy to be found in beautiful blooms, such as Mahonia. 183

JANUARY

One compost heap and 87 grass snake eggs were all successfully hatched; my contribution to the local herpetological community. 184

Observing and recording are integral parts of being a naturalist. 187

Most of these peas fail to make it as far as the kitchen. 189

A winter visiting fieldfare enjoying the apples I put out on the lawn. 190

A pike coming to an Alphonse "GT" fly in Bowood Lake Wiltshire. 191

Steeped in folklore, these 'elf cup' fungi can be found in late January if lucky. 194

Going, going gone! Photo credit Richard R Porter. 197

Photo credit Richard R Porter. 198

Photo credit Ian Harrison. 201

Long billed curlew Illustration from ca 1830. 203

FEBRUARY

Wiltshire Countryman Sid Vincent. 204

Habitats are the cornerstone to flourishing wildlife. This pond is in the middle of the Savernake Forest and with no nearby roads and no surrounding agriculture, the result is a thriving mass of frogs every springtime. 211

And then three weeks later the same pond, all the frogs are replaced with toads making an almost Canada goose-like chorus – what a place. 212

Recipe: Shepherd's Pie. 214

I asked David White, a superb wildlife photographer in Wiltshire, if he could get a picture of a robin on one of my spades. The robins were shy, so instead, here is a wonderful photo of blue-tit in action. 217

We only see siskins in the garden towards the end of February before they disappear again by March. 219

MARCH

David White set up a hide to take this picture in our front garden. 220

Steeped in folklore, these 'elf cup' fungi can be found in late January if lucky. 224

Potatoes nicely 'chitted' ready to plant out. 227

Forcing rhubarb accentuates the flavour and colours. 228

Chopped up it is like candy in a bowl. 229

My sea kale pots in the spring sunshine. 231

EPILOGUE

Having a pint outside a London pub after the book launch for Tarquin Millington-Drake's 'Living with Greys'. 232

A grey partridge in the North Pennines. Photo by Tarquin Millington-Drake. #whymoorlandsmatter. 233

I was inspired by nature as a child by the simple but beautifully illustrated Ladybird books. 235

We have many famous naturalists in the UK. 238

Photo credit Action for River Kennet – ARK. 242

After the sea trout 2am. 251

ABOUT THE AUTHOR
JOHN MILLER

John Miller is a passionate countryman, naturalist, and lifelong observer of the British countryside. Raised in a small Oxfordshire village, his deep love for nature began in childhood, where he was left to roam, a fairly feral child, where the fields were his playroom.

John's varied career has taken him across the globe, but his connection to the land and its seasonal rhythms remains the heart of his life and writing.

A premium writer on Scribehound, Board Member of the Atlantic Salmon Trust, and the Curlew Action charities, John is committed to conservation and sustainability. He is actively trying to push forward the agenda of the introduction of the new GCSE for natural history onto the syllabus, showcasing his dedication to education and environmental advocacy.

Through his writing, John encourages readers to open their eyes, be curious, and enrich their lives by reconnecting with nature. His work combines personal reflection with practical insights, aiming to inspire a greater awareness of the natural world and its importance in our lives. His enthusiasm for field sports, gardening, and sustainable living shines through in his stories, inviting others to explore the beauty and intricacies of nature.

In What's in a Year..., John blends personal anecdotes, topical lifestyle issues, and the current affairs of the natural world. His love letter to the countryside offers not only a seasonal journey but also lessons in curiosity and conservation, encouraging readers to ask the right questions when it comes to conservation and where applicable make small changes that can have a big impact on the environment.

After the sea trout 2am.